ANALOG/LOGIC COMPUTER
PROGRAMMING AND SIMULATION

ANALOG/LOGIC COMPUTER PROGRAMMING AND SIMULATION

Fred J. Ricci

HRB-SINGER, INC.
Reston, Virginia
and
Assistant Professor/Lecturer
The George Washington University
Washington, D.C.

SPARTAN BOOKS

NEW YORK • WASHINGTON

Copyright © 1972 by Spartan Books

All rights reserved. This book or parts thereof may not be reproduced without permission from the publisher.

Library of Congress Catalog Card Number 70-141371
ISBN 0-87671-166-2

Printed in the United States of America.

Sole distributors in Great Britain, the British Commonwealth, and the Continent of Europe:

        The Macmillan Press Ltd.
        4 Little Essex Street
        London WC2R 3LF

PREFACE

THE SIMULATION OF ELECTRICAL, Biological, Mechanical and Physical phenomena in general has taken on new and interesting proportions since the first attempts at simulation were made. In the past years, analog computers have played a major role in the simulation of physical systems. Digital computers as well have played a large part in the simulation of physical systems. However, only recently has the use of the analog with parallel logic become an important tool in the simulation field.

This combination of text book and work book is intended to present the fundamentals of Analog/Logic computer programming and simulation to beginners in the field, and to be a working reference to academic and industrial users of the computer and associated techniques. The book is designed to teach programming as well as to present pertinent fundamental problems and a comprehensive perspective of the theory, mechanization, and application of Analog/Logic computers in the simulation of physical systems.

The idea for the procedure of carrying out the experiments, which are in the Appendix of the book, was first conceived while I was teaching a computer programming course. An attempt has been made to include material from the various fields of Engineering so that this book may be used in an interdisciplinary fashion. Problems in the fields of Electrical, Mechanical, Chemical, Control Systems and Bio-Engineering have been included. In this way, the student will have an opportunity to practice his programming skills, utilizing problems from a vast range of Engineering disciplines. The intent here is to make the analog computer a part of the engineer's repertoire of problem solving mechanisms.

The text does not consider hybrid computers in the sense of large scale digital computers. However, the material does review one portion of a modern hybrid computer laboratory which consists of Analog/Logic and a general purpose digital computer.

This book proved to be a good aid to students studying in a course called Computer Methods. As part of this course the textual material in Chapters I through V was studied supported by problems A and F of the Appendix, and the Automobile Suspension System problem. The

PREFACE

covering of this material proved to be sufficient to give the student an introduction to Analog Computation.

This text was sufficient for a course in which approximately one-third was devoted to analog computation and the remainder to digital computation. In a one term undergraduate course in analog computation Chapters I through V could be covered in addition to about five problems in the Appendix. A graduate course given in the same subject would require the use of the entire book.

This book is intended to be used in many diversified ways and is more practical and engineering oriented than theoretical. No attempt has been made to include problems at the end of every chapter. The selection and implementation of problems is left up to the particular instructor. Any attempt by this author to include "set problems" at the end of every chapter would be purely presumptuous and would not lead to flexibility or help in keeping up with the state of the art.

It is hoped that this book will be of value not only to teachers wishing to have a text on Analog/Logic computers, but also to Engineers who would like to learn more about computers. The material is arranged so that all one needs is an analog computer and a desire to learn. It is recommended that a computer laboratory be used in conjunction with this text.

I would like to extend my thanks to the members of Electronic Associates for their technical assistance, to my wife Mary Jo and members of my family for their encouragement, and to Miss Claire Anderson of Monmouth College for her secretarial assistance.

—Fred J. Ricci

CONTENTS

Preface		v
List of Appendices		ix
I.	Analog/Logic Computer Programming and Simulation	1
II.	Analog Computer Linear Components	5
III.	Programming Analog Computers	22
IV.	Amplitude, Time Scaling and Checking Procedures	31
V.	Linear Systems Analysis	51
VI.	Non-Linear Programming Elements and Circuits	65
VII.	Analog/Logic Techniques and Interfacing (Hybrid Computer)	96
VIII.	Parallel Logic Components and Hybrid Programming	104
Index		229

APPENDICES OF
ANALOG COMPUTER PROBLEMS

A.	Spring Mass Damper System	129
B.	Tubular Chemical Reactor Control System	135
C.	Analysis of Tapered Nozzle	147
D.	Electron Ballistics	160
E.	D.C. Servo Simulation	173
F.	A CO_2 Rebreathing System	187
G.	Analysis of Kinetics Data	198
H.	Dynamic Behavior of Enzyme Systems	210
I.	R-L-C Transducer	220
J.	Estimation of Maximum Values From Two Simultaneous Differential Equations	226

CHAPTER I

Analog/Logic Computer Programming and Simulation

MANY MACHINES HAVE been developed which attempt to aid man in understanding nature. Digital computers are an aid which reduce problems to numerical techniques. These machines are accurate, have large memory capacities, operate very fast and can be very reliable. However, they do have limitations in that calculations are discrete and must be reduced to numerical techniques.

Analog computers also have their limitations, however they come much closer to approximating nature than digital computers. Analog computers are continuous, parallel machines which give an analogy of nature using electrical components.

They are the kind of machines which attempt to solve man's mathematical approximation of nature, namely differential equations in the most accurate way possible.

Analog computers are very flexible in their solution to differential equations in that they solve linear as well as non-linear equations and approximate the other non-linearities of nature very well.

The general purpose of an analog computer is an assembly of electronic and electromechanical components which, individually using d.c. voltages as variables, can perform specific mathematical operations. The equations most suitable for solution on such a computer are ordinary differential equations.

The sequential operation and flow of information in an analog computer study takes on a very specific form. It must be stressed that analog computers are intended to simulate physical phenomena and in doing so take on this very specific form of operations.

Figure I-1 indicates the kinds of procedures to be followed in setting up an analog computer to simulate some physical system or phenomena.

It should be recognized that the human operator is an integral segment

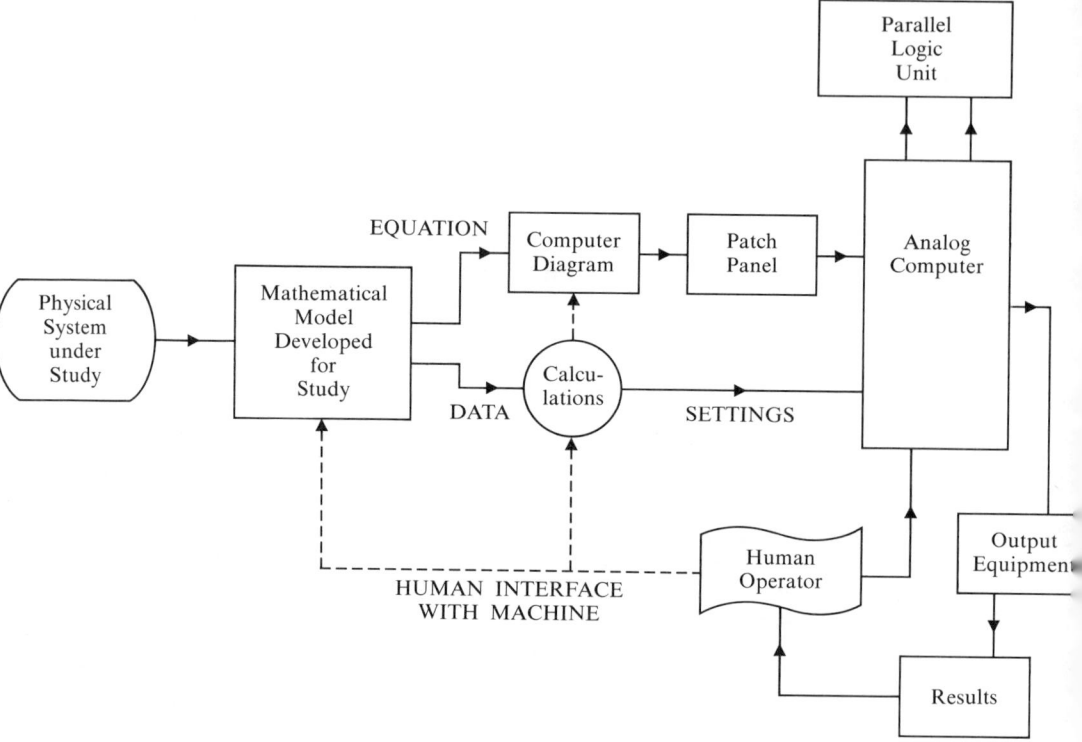

Fig. I-1. Interface between Analog Computer and Human Operator

of the entire closed loop system and offers the ingenuity and experience for coordinating the whole system.

Figure I-2 is a block diagram indicating the kinds of physical systems the computer can solve. It is important to point out that the solution to these equations depends as much on the programmer as it does on the computer. However, no matter how ingenious a programmer is he can never make up for the lack of or the diversity of equipment. There are many problems, especially very large, non-linear and partial differential equation problems, which require electronic logic apparatus to solve the problem accurately and with the greatest speed. Employing logic interconnected with the analog computer one will have a parallel logic computer or Analog/Logic (Hybrid) computer. This kind of computer may be implemented by adding logic components consisting of gates, flip-flops, registers, clocks of various speeds and trunks between the analog and parallel logic computer. The facility gained by adding this equipment greatly increases the capability of solving problems. Figure I-1 indicates the logic unit connected to the analog computer.

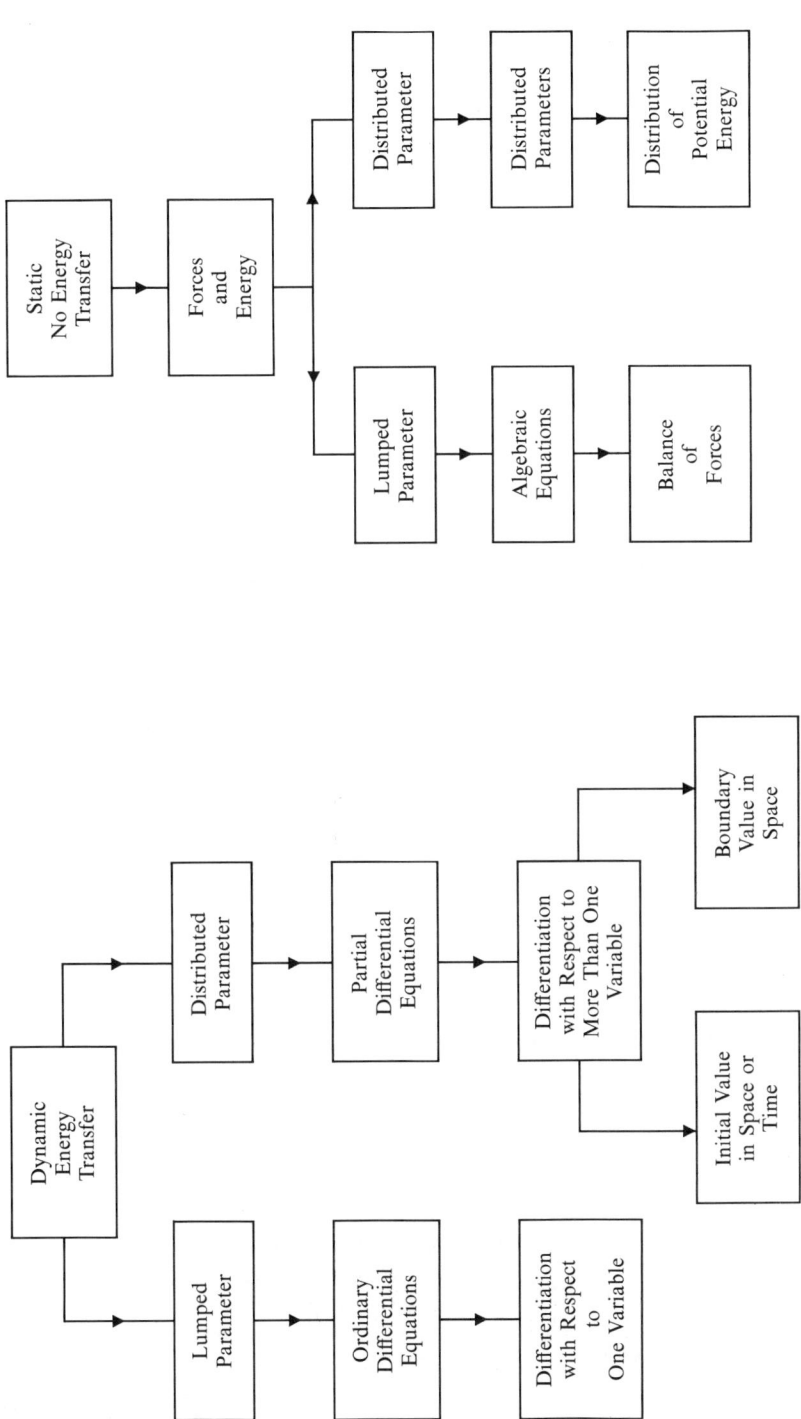

Fig. 1-2. Physical System That Can be Solved on Computer

The remaining chapters will discuss the components and techniques of analog computation, the interface components and finally the parallel logic components and their use. It is hoped that one will gain a facility for programming and using Analog/Logic (Hybrid) computers from this text.

The above discussion has been of a general nature. However, for a selected bibliography of computer applications consult the references.

CHAPTER II

Analog Computer Linear Components

ANALOG COMPUTERS ARE electromechanical devices that use many types of components in performing their electrical analogies. The components may be linear or non-linear depending on the function they serve. Before the actual components are considered it may be well to discuss the general characteristics of analog computers.

Analog computers come in many sizes and shapes. The basic differences, however, lie much deeper. Computers differ also in

(1) capacity (the number of computing components)
(2) capability (the quality of computing components and the operations they perform)
(3) reference voltage level (the operating voltage range of the computer). Typical voltages are:
$$\pm 10 \text{ or } \pm 100 \text{ volts}$$
(4) convenience factors (operator control, the accessibility of equipment, and others, some of which are less meaningful).

Modern analog computers are equipped with removable patch panels which contain the input, output and control terminations of the various analog components in a computer system. The input and output terminations of the components are connected in a particular configuration which is defined by the problem being solved. The control termination connections depend upon the mathematical operation required of a particular component, and the manufacturer's method of implementing the operating principles of that component. The term "patching" refers to the inter-connection of patch panel termination.

Specific patching information for a particular computing system can be obtained from the manufacturer's reference handbook for the computer.

It was mentioned earlier that analog components are classified as either linear or non-linear. The linear components perform the mathematical operations of

(1) multiplication by a constant,
(2) inversion,
(3) algebraic summation, and
(4) continuous integration.

These operations are sufficient to solve linear differential equations with constant coefficients of the general form

$$a_n\left(\frac{dx^n}{dt^n}\right)^m + a_{n-1}\frac{dx^{n-1}}{dt^{n-1}} + \cdots a_o X = f(t) \tag{1}$$

where all of the a_n's are constants and all of the derivatives have a degree $m = 1$.

Equations in which $m \neq 1$ and the a_n coefficients are not constants are considered to be non-linear differential equations. Non-linear components are used to solve these equations. The mathematical operations performed by non-linear components are:

(1) multiplication and division of variables,
(2) the generation of arbitrary functions, and
(3) the mechanization of constraints and elementary logic operations.

These components, together with the linear components, permit the analog computer to simulate the non-linear systems which occur in practice.

A consideration of the behavior of linear components will be considered first. The computer symbols for such devices are shown below.

(1) Potentiometer

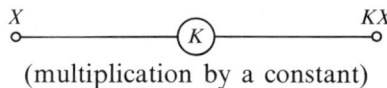

(multiplication by a constant)

(2) Inversion

(3) Summation

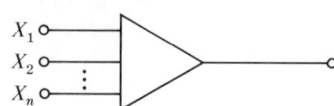

$$Y = -(X_1 + X_2 + \cdots X_n)$$

(4) Integration

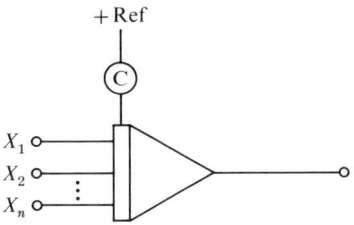

$$Y = -\int(X_1 + X_2 + \cdots X_n)\,dt - C$$

1. Potentiometers: Sometimes called attenuators, perform a multiplication of a d.c. voltage by a positive constant (K) which is less than unity. This device which is simply a fixed resistor with a movable wiper arm is shown in Fig. II-1.

Two types of pots, "grounded" and "ungrounded," are used in modern analog computers. These names are derived from the termination at the bottom, or "Lo" end, of the pot, as shown in the figure. The total resistance of a pot is of the order of 2,000 to 30,000 ohms and depends on the design of the computer.

Grounded potentiometers are used in conjunction with a reference voltage (a constant voltage source equal to the upper limit of the computer's operating voltage range) to obtain a fixed voltage less than reference voltage, or to multiply a problem variable by a constant less than unity. The input to the potentiometer is applied at its top or "Hi" end, the resultant output is obtained through the wiper arm.

Figure II-2 shows programmer symbols for both grounded and ungrounded pots. The ungrounded pot has special applications in addition to the attenuation of two variables, indicated in Fig. II-2 which will be discussed in later chapters.

Normally, an analog computer will contain one and one-half as many pots as it has amplifiers, and 80% of these will usually be grounded.

a. Loading and Setting of Attenuators: The potentiometers shown in Fig. II-2 are "unloaded" which means that no current is being drawn through the wiper arm (i.e. they are feeding an infinite resistance-open

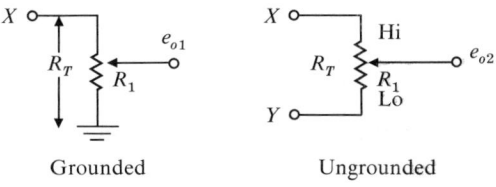

Fig. II-1. Schematic of Potentiometer

8   ANALOG/LOGIC COMPUTER PROGRAMMING AND SIMULATION

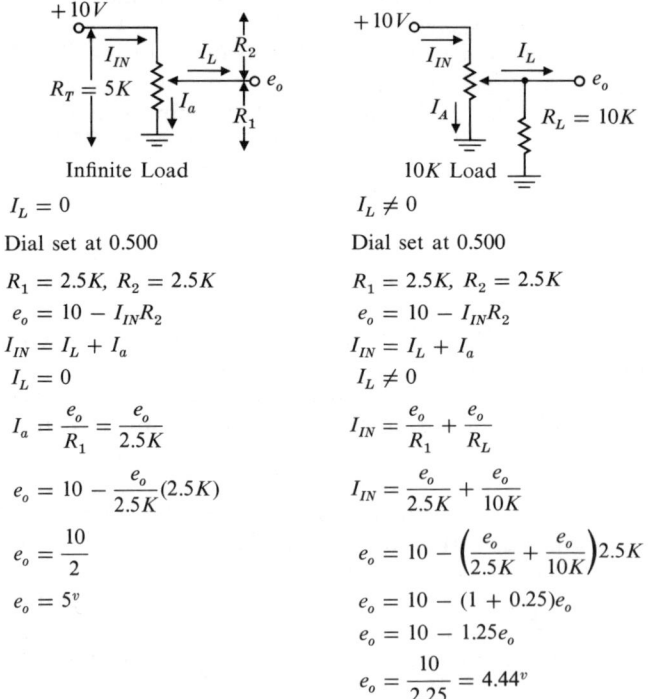

Fig. II-2. Programmer's Symbols for Both Grounded and Ungrounded Potentiometers

circuit). Therefore, the mechanical ratio, R_1/R_T, which can be set by a calibrated dial, is equal to its electrical ratio, e_o/e_{in}. However, this is not the case when the infinite load is replaced by a finite load as shown in Fig. II-3.

In practice, the wiper arm of a pot will be "looking into" a load ranging from 10^3 to 10^6 ohms since a potentiometer generally feeds resistor inputs to operational amplifiers. The effect of a $10K^*$ resistive load on a $5K$ pot set at 4/5 is shown in Fig. II-3.

Infinite Load

$I_L = 0$

Dial set at 0.500

$R_1 = 2.5K, R_2 = 2.5K$

$e_o = 10 - I_{IN}R_2$

$I_{IN} = I_L + I_a$

$I_L = 0$

$I_a = \dfrac{e_o}{R_1} = \dfrac{e_o}{2.5K}$

$e_o = 10 - \dfrac{e_o}{2.5K}(2.5K)$

$e_o = \dfrac{10}{2}$

$e_o = 5^v$

10K Load

$I_L \neq 0$

Dial set at 0.500

$R_1 = 2.5K, R_2 = 2.5K$

$e_o = 10 - I_{IN}R_2$

$I_{IN} = I_L + I_a$

$I_L \neq 0$

$I_{IN} = \dfrac{e_o}{R_1} + \dfrac{e_o}{R_L}$

$I_{IN} = \dfrac{e_o}{2.5K} + \dfrac{e_o}{10K}$

$e_o = 10 - \left(\dfrac{e_o}{2.5K} + \dfrac{e_o}{10K}\right)2.5K$

$e_o = 10 - (1 + 0.25)e_o$

$e_o = 10 - 1.25e_o$

$e_o = \dfrac{10}{2.25} = 4.44^v$

Fig. II-3. Potentiometer Loading

---

*In these notes, the following notation is used:

$K = 10^3, M = 10^6, m = 10^{-3}$ and $\mu = 10^{-6}$

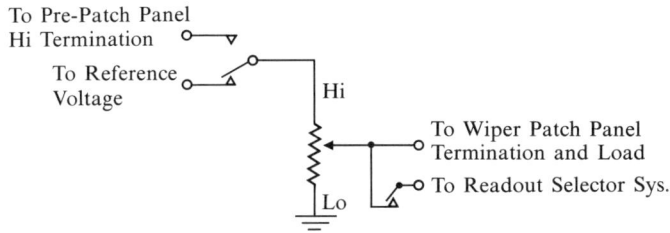

Fig. II-4.

In order to eliminate the effects of loading, potentiometers are set by monitoring the wiper voltage while the pot is "feeding" its normal load. In this way, it is possible to set potentiometers to three or four places depending upon the precision of the monitoring device.

In most computers, each potentiometer has switching associated with it similar to that shown in Fig. II-4.

When the switch is thrown, the patched input to the pot is replaced by a reference voltage, and the loaded wiper arm is connected to a monitoring device via a readout selector system. The readout device can be either a high impedance, digital voltmeter (DVM) *or* a null meter.

The more sophisticated analog computer systems have digitally-set attenuators. Here, the potentiometer is selected through a push-button system, and then is set by a servo device also controlled by push buttons.

2. Operational Amplifiers: The operational amplifier is the basic unit in the analog computer. It can be used in a "summing mode" to perform any or all of the three linear operations: inversion, summation and multiplication by a constant. It also can be used in an "integrating mode" to integrate a voltage or the sum of a number of voltages with respect to time.

Analog computer programs for investigating the behavior of physical systems require some operational amplifiers to be used as integrators, while others are used as "summers," "inverters," "high gain amplifiers," or in conjunction with special networks to perform nonlinear operations. Therefore, it is not necessary for all of the amplifiers to perform as integrators. In modern analog computers, a typical amplifier breakdown would be

(1) combination amplifiers capable of performing integration, summation or inversion . . . 30%
(2) summing amplifiers capable of performing summation and inversion . . . 45%

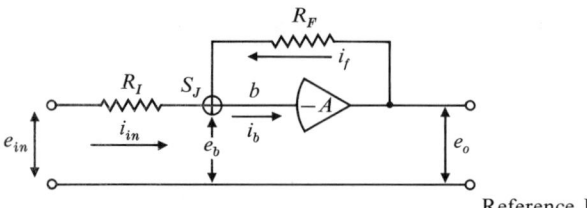

Fig. II-5. Simple Amplifier Circuit

(3) inverting amplifiers, capable of performing inversion only ... 25%

a. *Inversion and Multiplication by a Constant:* To understand the principle of the operational amplifier, consider the circuit* where a high gain d.c. amplifier (gain $= -A$) has a feedback resistor, R_F, and an input resistor, R_I (Fig. II-5). The d.c. amplifier is designed so that

(1) the amplifier output, e_o, is related to the summing junction voltage e_b, by the gain of the amplifier (i.e., $e_o = -Ae_b$ within the reference voltage range of the computer),
(2) the amplifier draws negligible current, $i_b \simeq 10^{-9}$ amps, and
(3) the gain of the amplifier is extremely high, usually on the order of 10^8 at d.c.

Using Kirchhoff's laws, the nodal current equation at the summing junction, SJ, is

$$i_b = i_{in} + i_f$$

or, from Ohm's law,

$$i_b = \frac{e_{in} - e_b}{R_I} + \frac{e_o - e_b}{R_F}.$$

Since $i_b \simeq 0$, it can be neglected. Replacing e_b by $e_o/-A$ we obtain

$$\frac{e_{in}}{R_I} + \frac{e_o}{AR_I} = -\frac{e_o}{AR_F} - \frac{e_o}{R_F}$$

$$e_o = \frac{-\frac{R_F}{R_I} e_{in}}{1 + \frac{1}{A}\left(\frac{R_F}{R_I} + 1\right)}.$$

---

*In this circuit, the input, e_{in}, summing junction, e_b, and output voltages, e_o, are referred to a reference level, such as ground. However, in the interest of simplicity, future circuits will omit the reference level terminal and consider it to be grounded. The gain of the amplifier, $-A$, will also be omitted in future circuit diagrams.

Since the ratio of R_F to R_I usually is less than thirty, and A is much greater than 1,

$$e_o = -\frac{R_F}{R_I} e_{in}. \tag{2}$$

From this equation we can see a most important characteristic of the operational amplifier: *the input-output relationship is solely dependent on the ratio of the feedback to the input impedances (resistances)*.

Using this equation as a basis for discussion, some of the various uses of the operational amplifier can be illustrated.

When both resistors are of equal magnitude, R, the amplifier output voltage has the same amplitude as the input voltage but is of the opposite polarity. Thus, the mathematical operation of inversion is performed

$$e_o = -\frac{R_f}{R_I} e_{in} = -\frac{R}{R} e_{in} = -e_{in}.$$

If the resistors are not of equal magnitude, the result is multiplication of the input by a constant. For example, if R_F were $1M$ and R_I were $100K$,

$$e_o = -\frac{R_f}{R_I} e_{in} = -\frac{1M}{0.1M} e_{in} = -10 e_{in}$$

or, if the resistance ratio is inverted,

$$e_o = -\frac{e_{in}}{10}.$$

b. <u>Summation</u>: The addition of two parallel input resistors to the previous circuit, yields

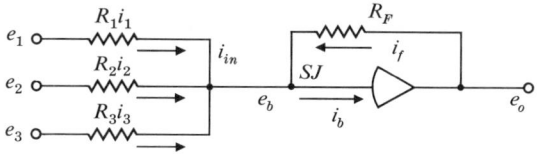

Fig. II-6. Summing Amplifier Circuit

And, the SJ node equation,

$$i_1 + i_2 + i_3 + i_f - i_b = 0.$$

Using Ohm's law, this equation becomes

$$\frac{e_1 - e_b}{R_1} + \frac{e_2 - e_b}{R_2} + \frac{e_3 - e_b}{R_3} + \frac{e_o - e_b}{R_f} - i_b = 0.$$

Since e_b and $i_b \simeq 0$, we have

$$e_o = -\left[\frac{R_F}{R_1}e_1 + \frac{R_F}{R_2}e_2 + \frac{R_F}{R_3}e_3\right]. \tag{3}$$

If the number of input resistors is increased to, say, N, the generalized summer equation becomes

$$e_o = -\left[\frac{R_F}{R_1}e_1 + \frac{R_F}{R_2}e_2 + \cdots + \frac{R_F}{R_N}e_N\right]. \tag{4}$$

c. <u>Integration</u>: When the feedback resistor used in previous circuits is replaced by a capacitor, the amplifier circuit for a single input becomes

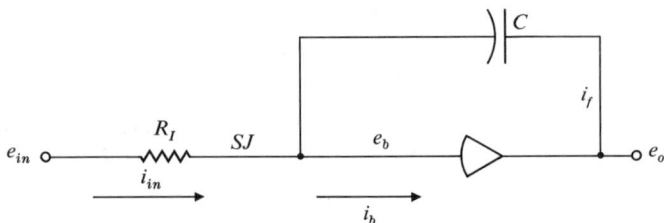

Fig. II-7. Simple Integrator Circuit

The relations among capacitance, voltage drop, and current for a capacitor with no initial charge is:

$$e = \frac{1}{C}\int_0^t i \, dt.$$

Thus, the voltage drop, $e_o - e_b$, across the feedback capacitor can be expressed as

$$e_o - e_b = \frac{1}{C}\int_0^t i_f \, dt$$

which can be differentiated to obtain an equation for i_f.

$$i_f = C\frac{d}{dt}(e_o - e_b).$$

A current summation at SJ ($e_b \simeq 0$, $i_b \simeq 0$) is now

$$\frac{e_{in}}{R_I} + C\frac{de_o}{dt} = 0$$

whose solution is

$$e_o = -\frac{1}{R_I C}\int_0^t e_{in} \, dt + \text{IC*}. \tag{5}$$

---

*IC refers to initial condition voltage on capacitor of amplifier.

We now have a device which can perform the operation of integration (with respect to time) on an input voltage.

For multiple resistor inputs, the integrator output is described by the equation:

$$e_o = -\int_o^t \left[\frac{e_1}{R_1 C} + \frac{e_2}{R_2 C} + \cdots + \frac{e_N}{R_N C} \right] dt + \text{IC}. \qquad (6)$$

It should be noted that the amplifier output voltage in this instance is the *integral of the algebraic sum of the input voltages.*

d. <u>Generalized Amplifier Equations</u>: If one defines the impedance, Z, of a passive element as

$$Z = \left| \frac{E}{I} \right|$$

where E is the voltage drop across the element, and I is the current passing through it, the input-output expression for the generalized circuit (Fig. II-8) is:

$$e_o = -\frac{Z_F}{Z_I} e_{in}.^*$$

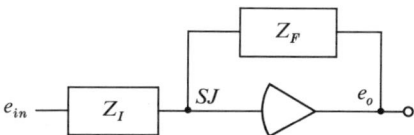

Fig. II-8. Generalized Amplifier Circuit

For a multiple input amplifier circuit

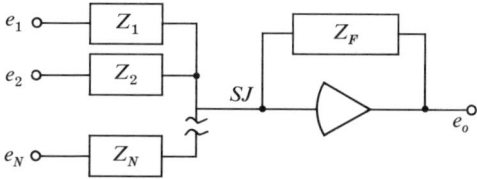

Fig. II-9. Generalized Multiple Input Amplifier Circuit

---

*For simplicity further reference to $e_i(t)$, $Z_i(t)$, etc. will be considered as functions of time and will be noted simply as e_i, Z_i, etc. unless indicated otherwise.

the input-output relationship is

$$e_o = -\sum_{n=1}^{n=N} \frac{Z_F}{Z_n} e_n = -\left[\frac{Z_F}{Z_1}e_1 + \frac{Z_F}{Z_2}e_2 + \cdots + \frac{Z_F}{Z_N}e_N\right]. \quad (7)$$

The impedance of a resistor is equal to its resistance in ohms

$$Z_R = R. \quad (8)$$

The impedance of a capacitor is time dependent. Recalling that the voltage drop across a capacitor is

$$e = \frac{1}{C}\int_0^t i\,dt$$

and defining the operators

$$p \equiv \frac{d}{dt} \text{ and } \frac{1}{p} \equiv \int_0^t dt,$$

the relation between voltage and current for a capacitor is

$$e = \frac{i}{pC}.$$

Since impedance is defined as the ratio of voltage drop to current the capacitor impedance is

$$Z_c = \frac{1}{PC}. \quad (9)$$

e. Programming Symbols: Before illustrating the programming symbols for the circuits just presented, it is important that one realizes how amplifiers and their associated passive elements are packaged in modern day computers. Each amplifier has associated with it an input network (resistors) and a feedback capacitor and/or resistor.

The input resistors are not equal in magnitude. Normally, one finds the input network containing from four to six resistors of two different magnitudes. For example, a six resistor input network may have three $0.1M$ and three $1M$ resistors.

The symbol used for a *high gain d.c. amplifier* is simply

An inverter

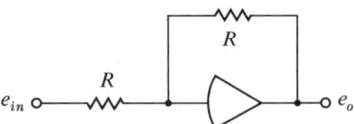

whose overall gain is unity because it has identical input and feedback resistors, is denoted by the programming symbol

If the passive elements were not identical the symbol would be

where G is the resistance ratio R_f/R_I.

In the case of summing amplifiers which can have multiple inputs, the programming symbol is

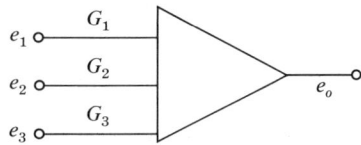

where $G_1 = R_F/R_1$, $G_2 = R_F/R_2$ and $G_3 = R_F/R_3$.

The symbol for an integrator, where $G = 1/R_I C$,

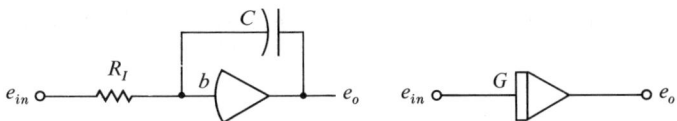

differs from that of a summer by a small rectangle which is *adjacent* to the base of the triangle. For multiple inputs, the symbol for an integrator becomes:

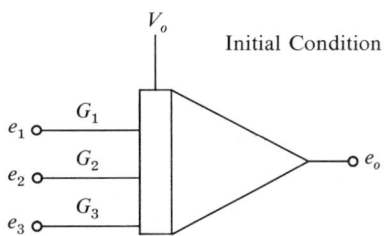

where $G_1 = 1/R_1C$, $G_2 = 1/R_2C$, and $G_3 = 1/R_3C$. The V_o input to the top of the integrator represents the initial value of e_o, or initial charge on the feedback capacitor, which will be discussed in the next section of this chapter.

Finally, one may have occasion to use a high gain amplifier with an input network but without a feedback element

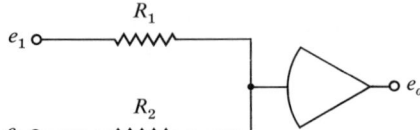

This is commonly represented by the symbol

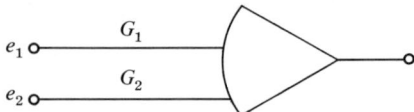

where G_1 and G_2 are inversely proportional to the size of the input resistors.

To coordinate the packaging of amplifiers and passive elements with the symbols just presented, it must be realized that the input terminations of the input networks usually are not labeled with the magnitude of the input resistors. They are labeled, rather, with gain factors which are based on standard feedback resistors and capacitors selected by the computer manufacturer for a specific computer system. For example, consider the input or patch panel terminations for an EAI TR-48 computer shown in Fig. II-10. Each input labeled 10 is a $10K$ resistor and each input

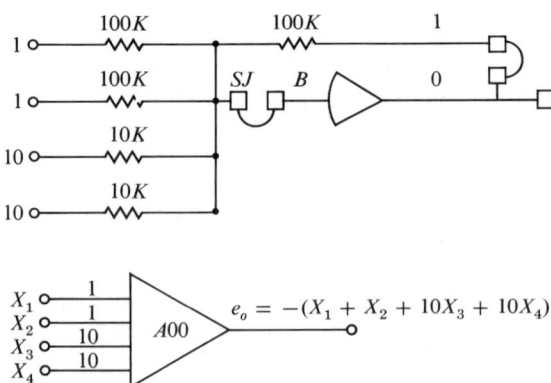

Fig. II-10. Summer Amplifier Patching

labeled 1 is a 100K resistor. Therefore, the *standard* feedback resistor for this system must be 100K.

It follows, then, that if this notation is to be used throughout this computing system, the standard integrating capacitor must be $10\mu f$ ($1/RC = 1$, 10 for 100K and 10K input resistors respectively). Patching details are a function of the specific computer.

3. Mode Control: The operator controls the mode of operation of the analog computer from pushbuttons or switches located on the control panel of the computer. The modes of computer operation are classified as follows:

a. Computational Modes
    Pot Set   (PS)
    Reset   (RS)
    Hold   (HD)
    Operate   (OP)
b. Check Modes
    Static Check (ST)
    Rate Test   (RT)

c. Slave Modes
    Slave   (SL)
    Tape   (TP)
d. Special
    Repetitive Operation (RO)

The Operate, Hold and Reset modes are the basic modes required for operation of an analog computer. The additional modes mentioned above are for operator convenience in large computing systems.

Analog computer control is accomplished by relays or electronic switches in the integrator circuitry. This is logical, since the integrator is the only dynamic element in the computer, that is, it is the only one which is time dependent. All other components are static in the sense that their output is directly related to their input at all times. A simplified schematic diagram (Fig. II-11) of an integrator includes input, feedback

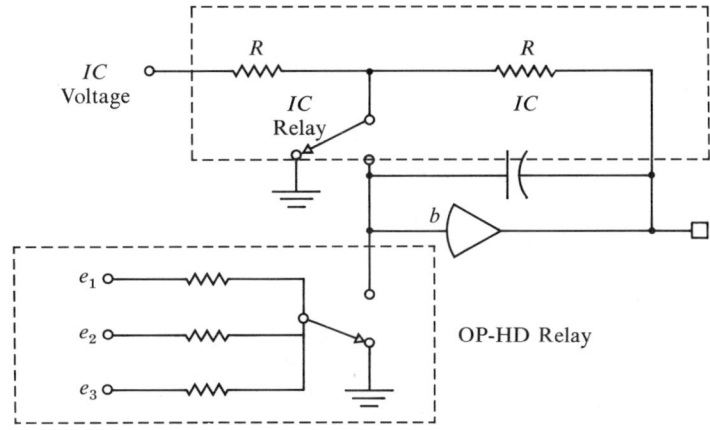

Fig. II-11. Simplified Integrator Control Diagram

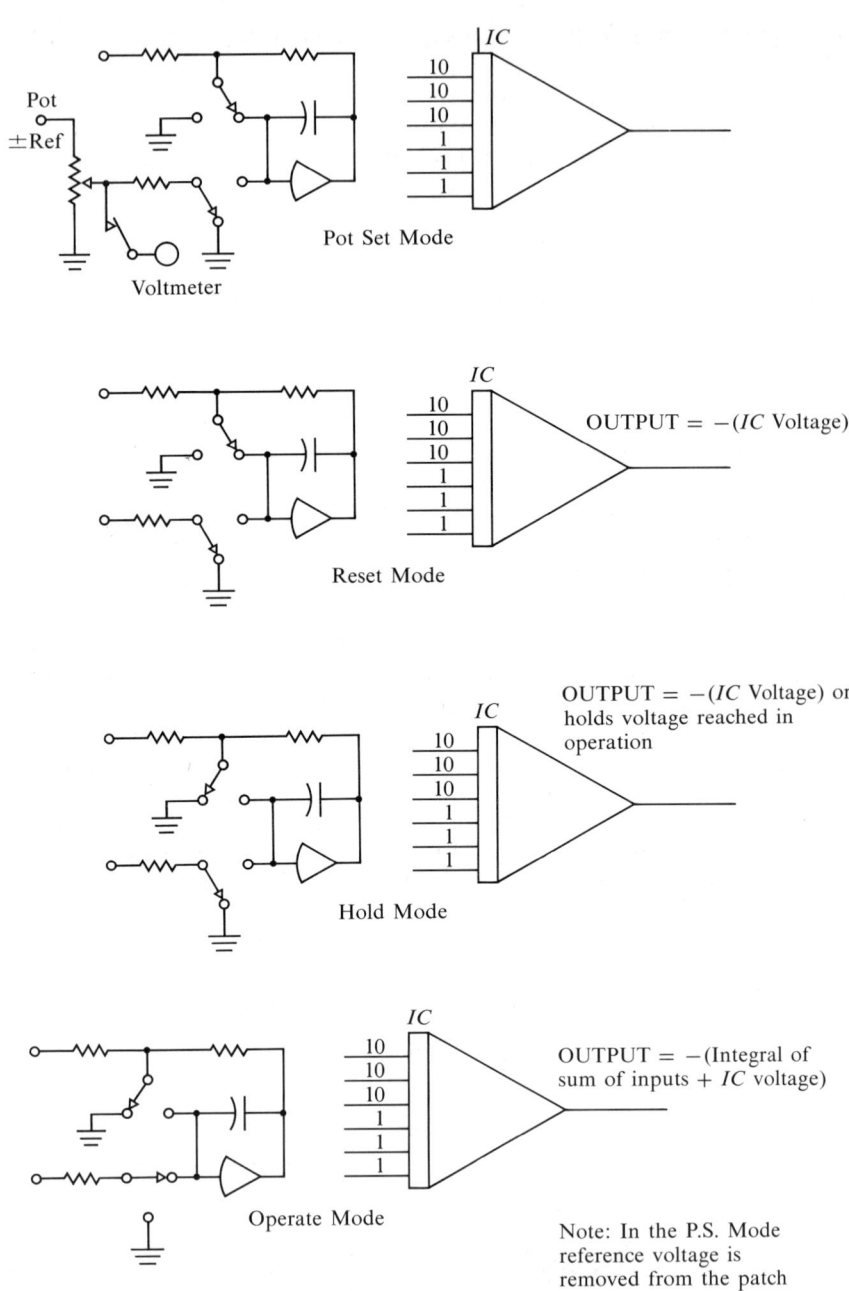

Fig. II-12. Integrator Mode Control

and initial condition (IC) networks, as well as Operate/Hold and Reset (IC) relays.

The IC relay is used by charging the integrator capacitor through the IC network to introduce an initial condition on the output of the integrator. The operate-hold (OP-HD) relay, in effect, starts and stops integration since problem variables cannot be summed and integrated unless the input network is connected to the amplifier. Note that this relay separates the inputs e_i to the amplifier from the input terminal, b, of the d.c. amplifier. Referring to Fig. II-12, consider the modes of operation and relay positions.

a. <u>Pot Set Mode</u>: To solve a problem on the analog computer the operator first must introduce the system parameters by setting attenuators. In most computers this is done in the pot set mode. However, in a few computers this function is performed in the reset mode. In the pot set mode, either the reference voltage terminals on the patch panel are de-energized, or the amplifiers are set to zero gain.

In pot set, the IC relay is connected to the input of the d.c. amplifier, and the summing junction of the input network is grounded through the OP-HD relay. This allows the attenuators to be set "looking into" their respective loads.

b. <u>Reset Mode</u>: The purpose of the reset mode is to introduce initial conditions on the integrator outputs. This can be done only if the IC voltage, e_{IC}, is connected to the amplifier; therefore the IC relay must change position. The OP-HD relay remains in the same position since problem solution is not required at this time. In the IC mode, the integrator circuit becomes

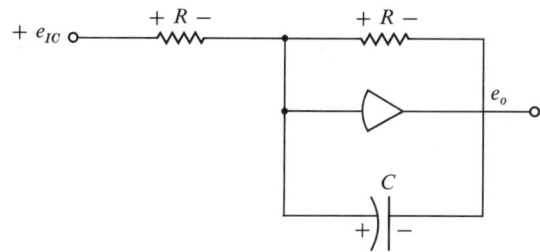

Fig. II-13

The input network is not shown as it is not physically connected to the amplifier in this mode.

The initial condition circuit is solving the equation

$$-\frac{de_o}{dt} = (e_{IC} + e_o)G = (e_{IC} + e_o)\frac{1}{RC}$$

where e_{IC} is a constant voltage and $G = 1/RC$. The solution of this equation is

$$e_o = -e_{IC}(1 - e^{-t/RC}) \tag{10}$$

indicating that the output voltage equals the *negative* of the initial condition voltage after $10RC$ time constants ($t = 10RC$, $e^{-10} \simeq 0$). This time constant is usually 0.1 seconds, ($G = 10$); therefore, $-e_{IC}$ appears at the amplifier output in less than one second.

c. Operate Mode: In the operate mode, the OP-HD relay closes and the integrator operates in the normal manner. In this mode, the IC relay must return to the ground position to remove the IC input, and the IC voltage does not affect the integrator output in the operate mode.

d. Hold Mode: In practice, it may be necessary to stop a problem solution, obtain intermediate results, and then complete the solution. Therefore, a capability of stopping the solution of a problem without destroying or losing e_o is desirable. This is accomplished by disconnecting the input network. Since the integrator input is zero, the amplifier output in hold is stored on the feedback capacitor, and will be constant.

The capacitors used in analog computers should be of very high quality to minimize "leakage" effects on the amplifier output voltage. (Leakage is equivalent to a high resistance across the capacitor.)

e. Static Test Mode: The static test mode is similar to the reset mode with one exception: special "reference voltage" terminations on the patch panel are energized in this mode only. Their purpose is to provide initial condition voltages for checking purposes for integrators, where initial-condition voltages are zero.

After a problem is mechanized on the computer, a static check (to be discussed in a later chapter) is performed to insure that the patching, etc., are correct.

A zero voltage is not a valid check, however, and, therefore, all integrator outputs must have initial condition voltages. If, as in the case of many analog computers, no static test mode exists, the static check can be performed in the reset mode. Here, one must connect integrator initial conditions to reference voltage sources physically and disconnect them after the check has been completed.

Check Amplifier

In performing a Static Check it is necessary that the net sum of the input voltages to all integrators be available, (i.e., the derivatives of variables). Consider an integrator in the reset mode:

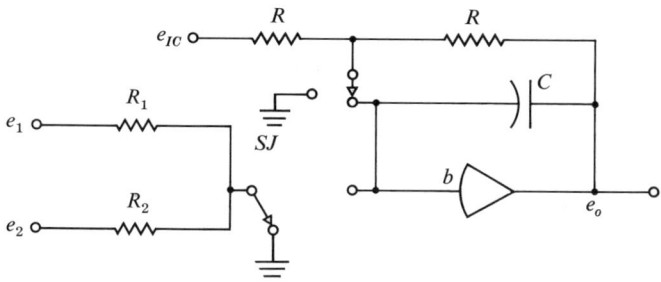

Fig. II-14. Integrator Circuit in the Reset Mode

The following operations must be performed to read the check point (derivative on an integrator):

(1) the integrator summing junction ground (through the OP-HD relay) must be removed,
(2) the summing junction must be connected to the base (b) of the check amplifier which has its own feedback resistor, and
(3) the output of the check amplifier must be connected to a voltmeter.

The implementation of these operations is either manual, which places the entire burden on the operator, or automatic, which is implemented by relays through a push-button selector system.

A choice of check amplifier feedback resistors is usually available to the operator by switching, so that he has a choice of feedback resistors.

This chapter has provided an introduction to analog computers, their linear components and controlling networks. The purpose here has been to present this material in the most general way possible so that the text will apply to future generation analog computers. In succeeding chapters the use of non-linear components and their characteristics will be discussed.

The next chapter will be concerned with programming analog computers using linear components.

CHAPTER III

Programming Analog Computers

THIS CHAPTER AND THE NEXT will be concerned with programming, scaling and checking of analog computer programs. Throughout the years many techniques have been utilized in teaching the concepts of the next two chapters. The technique used here is one of unit scaling. As the student is taught the various steps involved in programming, he will be able to try out his newly learned abilities by solving a problem related to the area learned. The analog computer solution to the automobile suspension system will be used as a common problem to demonstrate programming fundamentals.

The general problem of programming linear and non-linear differential equations on the analog computer involves some very basic steps. If the programmer follows these basic steps the solution to the problem will be quite straightforward; if he does not he may be in for some unpleasant surprises. A block diagram of the steps to follow is indicated in Fig. III-1.

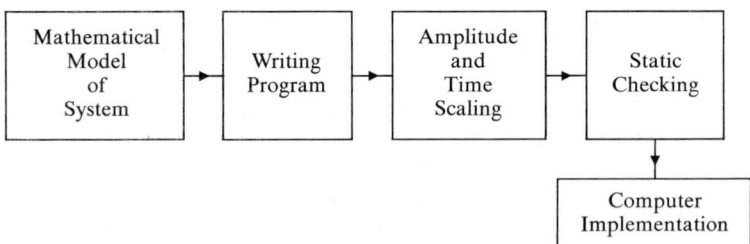

Fig. III-1. Programming Procedures

It can be seen from this diagram that the problem involves definition of some physical system in terms of differential equations and then manipulation of these equations on the analog computer using the potentiometer, integrator, summing amplifier, high gain amplifier and non-linear devices when desired.

Perhaps a simple example will demonstrate what is being mentioned. Let us start with a simple first order equation

$$\frac{dx}{dt} = -KX \ (K \text{ constant}). \tag{1}$$

This equation may describe a number of different physical phenomena, such as (a) radioactive decay, (b) dilution in a stirred tank, (c) the discharge of a capacitor, (d) first order chemical reaction, and many others.

The two variables in this equation are x and dx/dt. How can we represent the relation between them with analog components? Clearly, the component that accomplishes this is the integrator

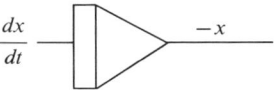

Fig. III-2

If dx/dt is fed into an integrator, the output will be $-x$. But where does dx/dt come from? From Eq. (1) we see that this is simply $-KX$. Since $-x$ is available at the integrator output, we can use a pot to multiply it by K.

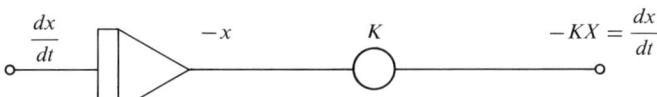

Fig. III-3. Analog Computer Circuit

This produces $-KX$ at the output of the pot. Since this is exactly what is needed at the integrator input, we can provide the desired input by connecting the pot output to the integrator input

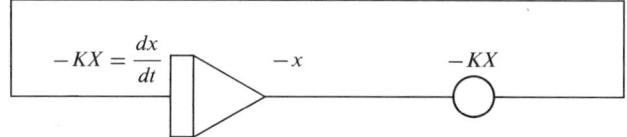

Fig. III-4. Solution to Differential Equation

In order to specify the solution completely, we need to know the initial value of x. (Equation (1) actually has infinitely many solutions, one for each initial condition.) If the initial condition is given, it may be implemented by using the circuit in Fig. III-5.

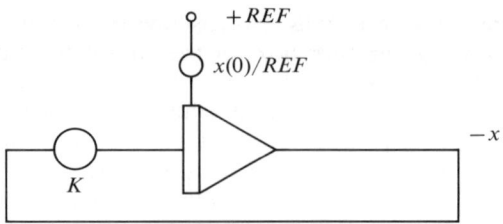

Fig. III-5. Complete Analog Computer Circuit

The programming technique used in solving the above problem is often referred to as the "bootstrap" method. We start by assuming that dx/dt (or $-dx/dt$) is available, and then integrate it to produce the variable x at the integrator output. We then use this integrator output to generate the derivative that we started with. The reasoning appears circular: If we have $-dx/dt$ available, then we can integrate it to obtain x; if we have x, we can multiply it by K to obtain $-dx/dt$. Does this "bootstrap" approach really lead to a valid solution of the given equation?

Perhaps the best way to indicate that it does is to consider the analogy between solving simultaneous algebraic equations and differential equations.

Consider the following two equations:

$$3y + 7z = 4 \qquad (2)$$
$$y + 2z = 1 \qquad (3)$$

solving for y and z:

$$y = \frac{4}{3} - \frac{7}{3}z \qquad (4)$$

$$z = \frac{1}{2} - \frac{y}{2}. \qquad (5)$$

It can be seen that the solution for y is dependent on z and the solution for z is dependent on y. Figure III-6 indicates a block diagram solution

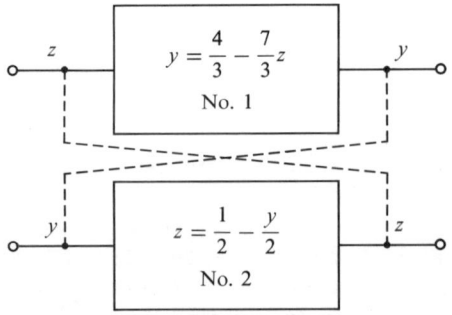

Fig. III-6. Solution to Simultaneous Algebraic Equations

for these simultaneous equations. Notice how the output of one box is needed as the input to another box for the complete solution.

This diagram demonstrates that in order to get a solution for z an input to y is needed and vice versa. Hence a solution to either equation would not occur unless the other was known. This is a "bootstrap" technique. The same technique can be utilized in solving the differential Eq. (1).

$$\frac{dx}{dt} = \dot{x} = -Kx. \tag{6}$$

If one writes another equation

$$\frac{dx}{dt} = \frac{dx}{dt}$$

or

$$x = \int \dot{X}\, dt \tag{7}$$

then a technique similar to the one used for the solution to simultaneous equations can be used.

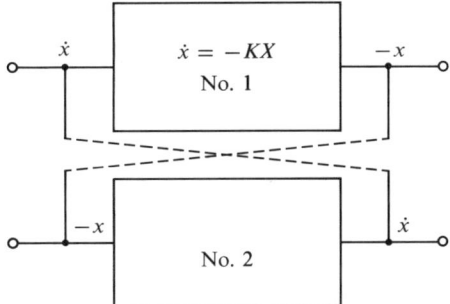

Fig. III-7. Solution to Differential Equations

If analog components are put in the boxes the solution is complete. It should be recognized that this solution is exactly equivalent to the one obtained in Fig. III-5.

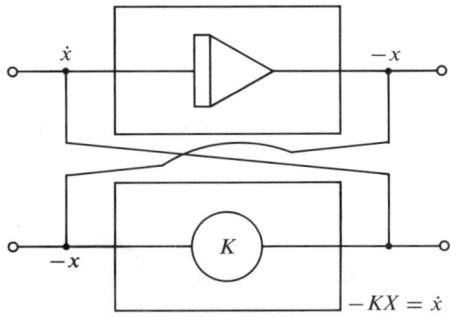

Fig. III-8

Perhaps the solution to another problem will demonstrate the techniques indicated above.

Consider the second order differential equation

$$a\ddot{x} + b\dot{x} + cx = 0 \tag{8}$$

where the dot over a variable represents differentiation with respect to time, that is,

$$\dot{x} = \frac{dx}{dt}.$$

A spring mass system with damping and an R-L-C circuit provide examples of physical systems described by this equation.

To solve this equation, first solve for the highest order derivative

$$\ddot{x} = -\left(\frac{b}{a}\right)\dot{x} - \left(\frac{c}{a}\right)x. \tag{9}$$

Then integrate this derivative to obtain lower order derivatives and the variables themselves. Feed these lower-order terms into the appropriate components, as called for by the equations, to generate the higher derivatives and close the loop.

Figure III-9 shows the computer diagram for the solution to this second order system.

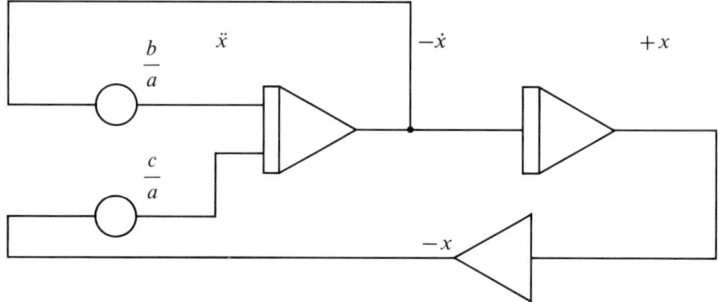

Fig. III-9. Computer Diagram for Solution to Second Order System

Now that the diagram has been provided the programmer must provide initial conditions from pots and reference voltages as required.

So far a first and second order system have been programmed. Suppose the equations become more complicated or there are more equations. Can general rules be written to solve any differential equation? The answer is yes. A summary of steps is indicated below.

Summary of Steps

(1) Obtain a description of the system to be studied in terms of ordinary differential equations.

(2) Solve each equation for the highest derivative that occurs in it.
(3) Integrate these derivatives to obtain the lower-order derivatives and the variables themselves.
(4) Feed these lower-order terms into the appropriate components, as called for by the equations, to generate the highest derivatives and "close the loop."
(5) Provide initial conditions from pots and reference voltages, as required.

Step (4) can be carried out in more than one way since there are generally many different but equally correct circuits for any problem.

Now that a background for programming has been given, a more complicated problem will be tried. The automobile suspension system problem will be used as an example in this chapter and the following to demonstrate the concepts of programming, scaling and static checking. The problem is divided up into sections so that one section at a time may be studied.

Automobile Suspension System

Introduction

The simulation of physical systems on the analog computer is one of its most powerful functions. By simulation techniques, one is saved the costly situation of building a system and changing the components for a parameter study. Simulation on the computer enables one to study a system, with all of its variations, without physically building it.

Coupled mechanical systems usually require careful study and experimentation to guarantee their correct behavior. This problem presents a particular type of coupled system, namely a simplified version of an automobile suspension system, for simulation on the computer. Although there are many variations of this system, the one to be studied will have all of the component values specified.

Problem Statement I

Consider the system in Fig. III-10, which is a simplified model of one wheel of an automobile suspension system. The spring action of the tire provides us with one equation and the action of the auto spring and shock absorber another. A force-balance yields the equations

$$M_1 \ddot{x}_1 + D(\dot{x}_1 - \dot{x}_2) + K_1(x_1 - x_2) = 0 \quad (10)$$
$$M_2 \ddot{x}_2 + D(\dot{x}_2 - \dot{x}_1) + K_1(x_2 - x_1) + K_2(x_2 - x_3) = 0. \quad (11)$$

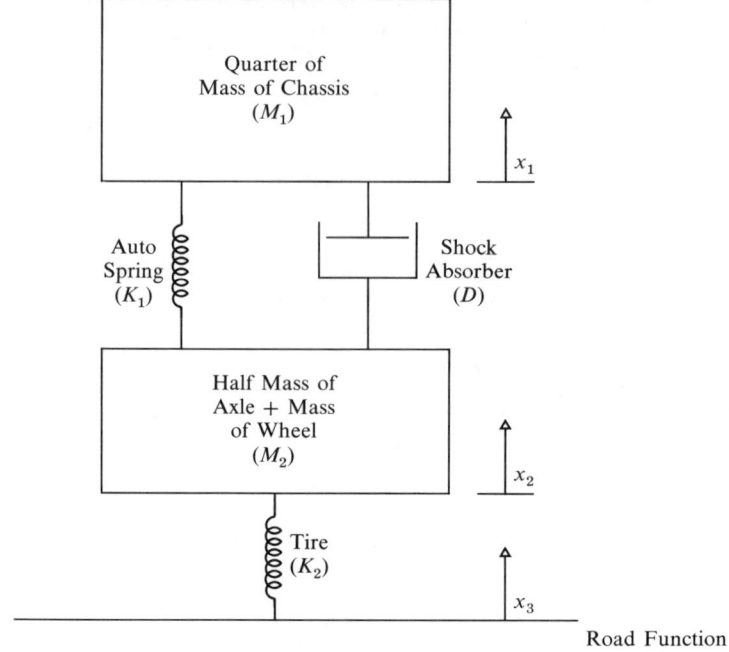

Fig. III-10

For simplicity, we may assume that x_3 is a step function (the car might be riding up onto a curb). Other more complicated functions are possible, of course. However, a step input keeps the program simple. The quantities K_1, K_2, M_1, M_2, D and x_3 are all constant for a given computer run. These equations should be programmed so that exactly eight (8) amplifiers are utilized. Hint: generate terms involving the velocity differences and displacement differences. The values of the parameters encountered are

$$M_1 = 25 \text{ slugs}$$
$$M_2 = 2 \text{ slugs}$$
$$x_3 = 5 \text{ in.} = 5/12 \text{ ft.}$$
$$20 \leq D \leq 200 \text{ lb./ft. per sec.}$$
$$K_1 = 1000 \text{ lb./ft.}$$
$$K_2 = 5000 \text{ lb./ft.}$$

SOLUTION TO PROBLEM STATEMENT I

Figure III-11 shows the unscaled computer diagram for the solution of Eqs. (12) and (13).

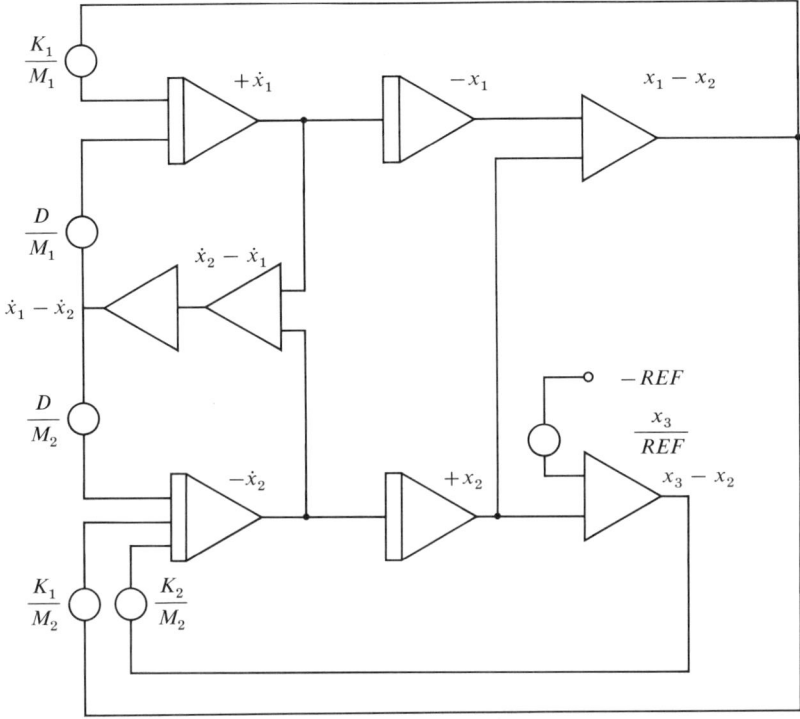

Fig. III-11

$$-\ddot{x}_1 = \frac{D}{M_1}(\dot{x}_1 - \dot{x}_2) + \frac{K_1}{M_1}(x_1 - x_2) \tag{12}$$

$$\ddot{x}_2 = +\frac{D}{M_2}(\dot{x}_1 - \dot{x}_2) + \frac{K_1}{M_2}(x_1 - x_2) + \frac{K_2}{M_2}(x_3 - x_2). \tag{13}$$

Note that each equation was solved for its highest order derivative, and that one equation was multiplied through by minus unity in order to meet the eight amplifier requirement. The amplitude, time scale and static checking will be considered in Chapter IV. Further programming examples may be found in Appendix I.

Conclusion

This chapter has attempted to demonstrate the fundamentals of analog computer programming from a practical point of view. First a very simple problem was solved and then the problems became more difficult until the student was actually given a problem to solve. It should be kept in mind that all of the problems solved used only linear components. In Chapter VI non-linear components will be considered.

The next chapter will concern itself with the scaling and checking of analog computer circuits.

CHAPTER IV

Amplitude, Time Scaling and Checking Procedures

JUST DRAWING A computer diagram is not enough. A programmer must also carry out the function of amplitude and time scaling in addition to checking procedures. This chapter will concern itself with these aspects in addition to practicing these functions on the automobile suspension system problem.

The need for amplitude and time scaling arises because of the incompatibility of the analog computer with natural phenomena. The analog computer can only vary between plus and minus one machine unit (plus and minus X volts), whereas natural phenomena may have amplitudes which are very large or exceedingly small. The analog computer may operate in one time frame, whereas natural phenomena take place over long periods of time or very short periods of time.

From an engineer's point of view, amplitude and time scaling are necessary so that all natural phenomena may be observed on the computer within the voltage limits of the machine and the time limits of man and the machine. Man is not capable of interpreting results that happen too quickly or not quickly enough.

In particular, if one wanted to display a sine wave that had a magnitude of 5000 lb. and a frequency $\omega = 1$ radians/sec., that is,

$$e(t) = 5000 \sin t$$

it would not be possible because the machine only goes to $\pm XV$ and the variations would be too large for man to interpret. Hence amplitude and time scaling would be necessary. That is, a compression of the wave would have to take place. Figure IV-1 illustrates the compression necessary for the proper amplitude and time scaling. Notice that curve (a) represents a sinusoidal wave within the limits of the machine and operating fast enough so that many variations are observable. Curve (b) is

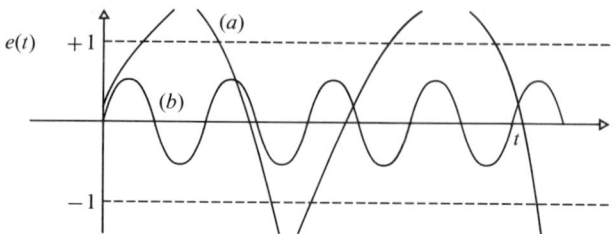

Fig. IV-1. Amplitude and Time Scaling of a Sinusoidal Wave

said to be *nicer* from an engineering point of view because one can obtain more information from it.

The author believes that the best way to teach amplitude and time scaling is to consider a specific example.

Consider the second order equation of Chapter III:

$$a\ddot{x} + b\dot{x} + cx = 0. \tag{1}$$

The physical system that this equation represents is

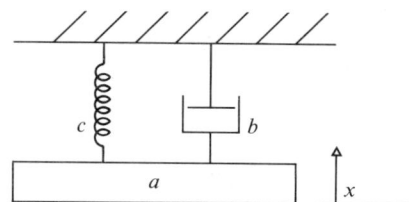

Fig. IV-2. Spring Mass Damper System

where: a—mass constant
b—dashpot constant
c—spring constant.

Physically this system will vibrate about some point with a displacement X and some natural frequency ω. The system may have very large or very small variations of displacement and may move very slowly or very rapidly. It should be pointed out that the displacement of the system is purely a function of the driving force and the behavior (qualitative) and vibratory properties are functions of the parameters of the system. These physical phenomena suggest that amplitude and time scaling are directly related to the physical properties of the system and the forces on it.

With these thoughts in mind let us consider Fig. IV-3 with parameter values at $a = 20$, $b = 100$ and $c = 60$, and maximum values for x and $\dot{x}$ at $|x|$ max $= 5$ cm and $|\dot{x}|$ max $= 50$ cm.

AMPLITUDE, TIME SCALING AND CHECKING PROCEDURES 33

Thus, in order to keep the variation of x within one machine unit, x would have to be divided by x max, then

$$\left|\frac{x}{x \max}\right| \leq 1$$

similarly,

$$\left|\frac{\dot{x}}{\dot{x} \max}\right| \leq 1.$$

Therefore, if these could be made outputs of amplifiers instead of $\dot{x}$'s and x's the amplitude problem would be solved. It should be stressed, at this point, that in amplitude scaling the outputs of all amplifiers must be kept between plus and minus one machine unit.

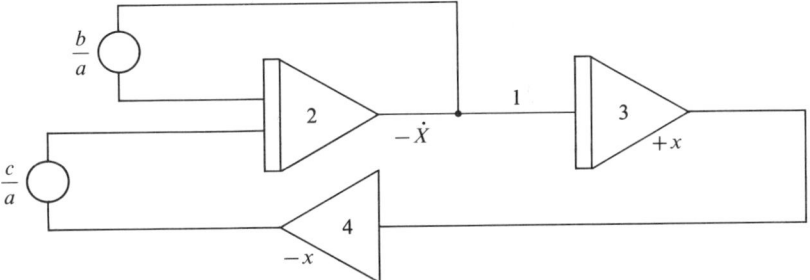

Fig. IV-3. Analog Computer Circuit for the Solution of Second Order System with All Parameters Isolated

If Eq. (1) is solved for the highest order derivative,

$$\ddot{x} = -\left(\frac{b}{a}\right)\dot{x} - \left(\frac{c}{a}\right)x \tag{2}$$

it can be observed that x and $\dot{x}$ are the outputs of amplifiers 2 and 3 respectively and should be kept between plus and minus one machine unit. Rewriting Eq. (2) to take into account maximum values without changing the equation; Eq. (3) results:

$$\frac{d}{dt}\left[\frac{\dot{x}}{\dot{x} \max}\right]\dot{x} \max$$

$$= -\left(\frac{b}{a}\right)\left[\frac{\dot{x}}{\dot{x} \max}\right]\dot{x} \max - \left(\frac{c}{a}\right)\left[\frac{x}{x \max}\right]x \max. \tag{3}$$

It should be noticed that the outputs of all amplifiers are now between plus and minus *one machine unit*. Solving Eq. (3) for $d/dt[\dot{x}/\dot{x} \max]$, Eq. (4) results

$$\frac{d}{dt}\left[\frac{\dot{x}}{\dot{x} \max}\right] = -\left(\frac{b}{a}\right)\left[\frac{\dot{x}}{\dot{x} \max}\right] - \frac{x \max}{\dot{x} \max}\left(\frac{c}{a}\right)\left[\frac{x}{x \max}\right]. \tag{4}$$

Notice that the outputs of all the amplifiers are now within brackets.

Rewriting Eq. (4) once more indicates very specifically the representation of the physical parameters on the computer.

$$\frac{d}{dt}\left[\frac{\dot{x}}{\dot{x}\ \text{max}}\right] = -\underbrace{\left(\frac{b}{a}\right)}_{\text{(Gain)}}\underbrace{\left[\frac{\dot{x}}{\dot{x}\ \text{max}}\right]}_{\text{(Pot Setting)}} - \underbrace{\left(\frac{c\ x\ \text{max}}{\dot{x}\ \text{max}\ a}\right)}_{\text{(Gain)}}\underbrace{\left[\frac{x}{x\ \text{max}}\right]}_{\text{(Pot Setting)}} \qquad (5)$$

Amplifier Outputs

The coefficients in front of the amplifier outputs are the product of the amplifier gains and potsettings. The amplifier outputs determine the *amplitude* scaling and the coefficients determine the *time* scaling. For example, Eq. (5) implemented on the computer is indicated in Fig. IV-4.

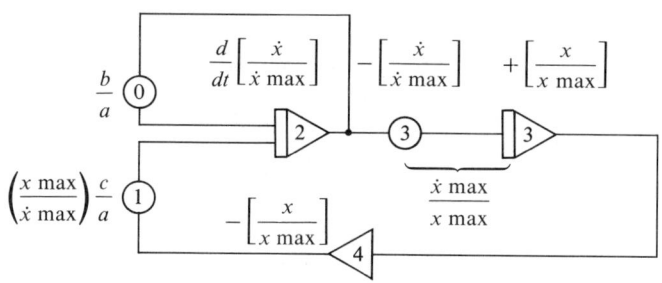

Fig. IV-4 Scaled Computer Diagram

It can be noticed that this circuit is different from that shown in Fig. IV-4 for a number of reasons. First of all, a potentiometer appears between amplifiers 2 and 3 (its use will be explained later), secondly no potentiometer settings or gains have been specified. The potentiometer settings and gains depend on the coefficients of the outputs on amplifiers indicated in Eq. (5). If one looks at each coefficient separately and considers its value for the maximum values of a, b and c given, values for potentiometer settings and gains may be determined:

$$\left(\frac{b}{a}\right) = \frac{100}{20} = 5$$

$$\left(\frac{c\ x\ \text{max}}{\dot{x}\ \text{max}\ a}\right) = \frac{60(5)}{(50)(20)} = 0.30$$

One must now choose a combination of potentiometer setting and amplifier gain such that these conditions can be met.

Since:

$$\left(\frac{b}{a}\right) = \frac{100}{20} = 5 = \overset{\text{AMPLIFIER GAIN}}{10}(\overset{\text{POTENTIOMETER SETTING}}{0.5})$$

$$\left(\frac{c\, x\, \max}{\dot{x}\, \max\, a}\right) = \frac{60(5)}{(50)(20)} = 1(0.30)$$

Figure IV-4 may now be redrawn with the appropriate amplifier gains and potentiometer settings. If it should occur that the coefficients appearing in the equation are not possible to obtain, that is, the product of potsetting and gain is not between 0.05 and 10, that is, $0.05 \leq$ (POT SETTING) (GAIN) ≤ 10 then *time scaling* is probably necessary. Time scaling will be considered in more detail in the following section.

Before the redrawing of Fig. IV-4 is considered, it may be well to explain the reason for the potentiometer between amplifiers A2 and A3. Amplifier A3 integrates the output of amplifier A2, therefore without potentiometer 3 the output of A3 would be

$$A3 = -\int_0^t (\text{output of A2}). \qquad (6)$$

or

$$A3 = -\int_0^t \left[\frac{-\dot{x}}{\dot{x}\, \max}\right] dt \qquad (7)$$

$$A3 = \left[\frac{+x}{\dot{x}\, \max}\right]. \qquad (8)$$

It can be seen that Eq. (8) does not agree with the output of A3 indicated in Fig. IV-4, that is,

$$\frac{x}{\dot{x}\, \max} \neq \frac{x}{x\, \max}. \qquad (9)$$

If however, potentiometer 3 is inserted, the output of A3 becomes:

$$A3 = -\left(\frac{\dot{x}\, \max}{x\, \max}\right)\int_0^t -\left[\frac{\dot{x}}{\dot{x}\, \max}\right] dt \qquad (10)$$

and

$$A3 = +\left[\frac{x}{x\, \max}\right]. \qquad (11)$$

Equation (11) now agrees with the output of amplifier A3 and is thus properly scaled. The rule of thumb for programming then is: *A poten-*

tiometer must be placed between all integrators in an analog computer circuit to obtain a scaled diagram.

Figure IV-5 is a completely scaled computer diagram for the maximum values and parameters given. IT SHOULD BE NOTED THAT TIME SCALING WAS NOT NECESSARY IN THIS CIRCUIT.

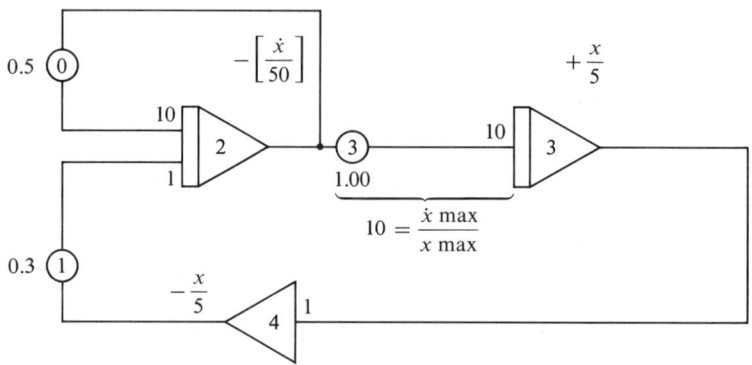

Fig. IV-5. Completely Scaled Computer Diagram

It should be pointed out, once again, that the coefficients of the differential equation are constant values for linear systems and are set on the analog computer by adjusting the product of the pot settings and gains of amplifiers.

To make the bookkeeping easier, especially for large problems, the following kind of table should be helpful:

TABLE IV-1. Table to Keep Record of Maximum Values

Problem Variable	Estimated Maximum	Computer Variable
$\dot{x}$	50 cm	$[\dot{x}/50]$
x	5 cm	$[x/5]$

At this point, one should have the ability to amplitude scale a differential equation and a feeling for why and when a differential equation should be time scaled. The succeeding pages will be concerned with the implementation of time scaling.

It was mentioned previously that time scaling arises because physical events either happen too quickly or too slowly for human operators to observe on the analog computer. Thus the computer must be slowed down or speeded up to allow things to happen in a reasonable amount of time, that is, many events of physical phenomena must be allowed to occur within a time that a human operator has a chance to completely observe the phenomena.

A technique that is usually accepted is to define some constant β, such that:

$$\beta = \frac{\tau}{t} = \frac{\text{Computer Time}}{\text{Problem Time}} \tag{12}$$

where β is the time-scale factor.

Thus β has the units of computer time (usually seconds) divided by problem time. If problem time is also measured in seconds, then β is dimensionless. In this case, the magnitude of β indicates the factor by which the problem is speeded up or slowed down. If $\beta > 1$, the computer solution is slower than the original process. For example, if $\beta = 7$, then, from Eq. (12), $\tau = 7t$. When t equals 1 second, τ equals 7 seconds, that is, an event that takes place in one second in the original problem requires 7 seconds on the computer. If $\beta < 1$, the computer solution is faster than the original process.

Now that this is understood the question is, how does one implement a time-scale change on the computer? Examination of the standard analog-computer components indicates that only one of them has anything to do with time—the integrator. For all other components, the relation between the inputs and outputs can be described without mentioning time; therefore all components other than integrators are unaffected by time-scaling.

To derive a relation for the input/output characteristics of an integrator refer to Fig. IV-6. What is needed is a device which integrates with respect to *machine* time τ, not *problem* time t.

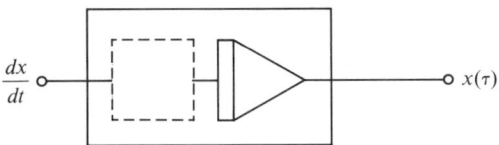

Fig. IV-6. τ-integrator Network

Essentially what is happening is that a time derivative is being fed into the integrator network and a variable as a function of τ is wanted on the output.

From Eq. (12) and the input to τ-integrator network one can develop an expression which will dictate the component to be placed in front of integrator.

Using,

$$\beta = \frac{\tau}{t}; \quad t = \frac{\tau}{\beta}$$

and

$$\frac{dx}{dt} = \frac{dx}{dt} \tag{13}$$

$$\frac{dx}{d(\tau/\beta)} = \frac{dx}{dt} \tag{14}$$

$$\beta \frac{dx}{d\tau} = \frac{dx}{dt} \tag{15}$$

or finally

$$\frac{dx}{d\tau} = \left(\frac{1}{\beta}\right)\frac{dx}{dt}. \tag{16}$$

Therefore if dx/dt can be multiplied by $1/\beta$, $dx/d\tau$ will result. If $dx/d\tau$ is integrated by an integrator $x(\tau)$ will result and the problem is complete. Figure IV-7 depicts the complete τ-integrator network. Notice the inclusion of a potentiometer set to $1/\beta$.

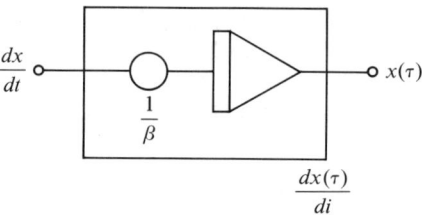

Fig. IV-7. Complete τ-integrator Network

Thus the following general rule is arrived at: to speed up or slow down the computer solution by a factor of β, *every integrator input must be divided by β*.

This factor $1/\beta$ is in addition to the voltage scale factors and other coefficients required by the problem. For example, consider Eq. (5) rewritten here.

$$\frac{d}{dt}\left[\frac{\dot{x}}{\dot{x}\,\text{max}}\right] = -\left(\frac{b}{a}\right)\left[\frac{\dot{x}}{\dot{x}\,\text{max}}\right] - \left(\frac{c\,x\,\text{max}}{\dot{x}\,\text{max}\,a}\right)\left[\frac{x}{x\,\text{max}}\right]$$

In order to time-scale this problem the t-variable must be changed to a τ-variable by

$$\frac{dx}{d\tau} = \left(\frac{1}{\beta}\right)\frac{dx}{dt}$$

then,

$$\frac{d}{d\tau}\left[\frac{\dot{x}}{\dot{x}\,\text{max}}\right] = -\left(\frac{b}{\beta a}\right)\left[\frac{\dot{x}}{\dot{x}\,\text{max}}\right] - \left(\frac{c\,x\,\text{max}}{\beta \dot{x}\,\text{max}\,a}\right)\left[\frac{x}{x\,\text{max}}\right]. \tag{17}$$

AMPLITUDE, TIME SCALING AND CHECKING PROCEDURES

This can be implemented on the computer in the following way (note Fig. IV-8).

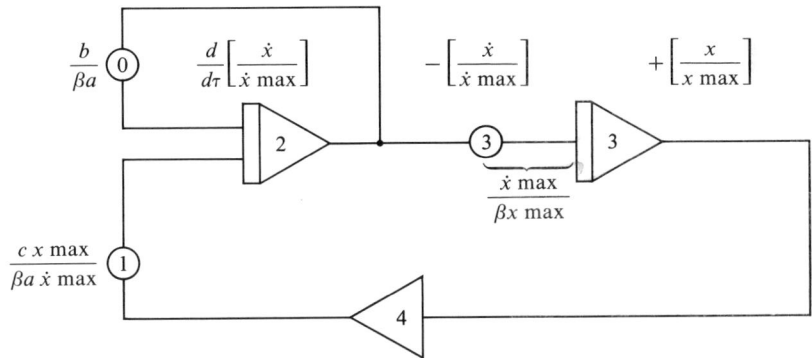

Fig. IV-8. Completely Scaled Computer Diagram

Figure IV-8 is now a completely *amplitude* and *time-scaled* computer diagram for the solution to the second order differential Eq. (1). The only thing left to consider is choosing the value of β for a particular problem. Since β is a time-scale factor its value is dependent upon the physical characteristics of a system. In this problem β is dependent on a, b and c. That is, one can choose β by looking at the coefficients of the amplifier output variables. If these coefficients are less than 0.05 or greater than 10 a value of β is needed, otherwise β is generally equal to one ($\beta = 1$). For example, if $a = 20$, $b = 100$ and $c = 60$ it was shown previously that the proper potentiometer settings and amplifier gains could be chosen without any further manipulation. However, if $a = 10$, $b = 1000$ and $c = 600$ then the coefficients of the untime-scaled problem become

$$\left(\frac{b}{a}\right) = \frac{1000}{10} = 100$$

$$\left(\frac{c\,x\,\max}{\dot{x}\,\max\,a}\right) = \frac{600}{(50)(10)} = 6.0.$$

Since these values are not between 0.05 and 10 they can not be gotten on the computer. Now, if the time-scaled coefficients are considered, one can see that β can be chosen to get the coefficients in the proper range. *This is time scaling:*

$$\left(\frac{b}{\beta a}\right) = \frac{100}{\beta}$$

$$\frac{c\,x\,\max}{\beta(\dot{x}\,\max)a} = \frac{6.0}{\beta}$$

if $\beta = 10$, this value will allow one to obtain the coefficients on the computer, that is, for $\beta = 10$:

$$\frac{100}{\beta} = 10$$

$$\frac{6.0}{\beta} = 0.6.$$

Certainly one can think of products of *pot settings and gains* which will enable one to obtain these values.

To get a better feeling and deeper insight into the problem of amplitude and time scaling in addition to becoming acquainted with static checking procedures study the continuation of the automobile suspension problem which follows.

Problem Statement II

The automobile suspension system, which was programmed in Chapter III without scaling, should now be both amplitude and time scaled.

The parameters have the values (or ranges) indicated in problem statement I, Chapter III. Under these conditions, the maximum values of x_1 and x_2 will be approximately one foot each. This problem should be studied very carefully so that one will completely understand *amplitude, time scaling* and *static checking procedures*.

Solution to Problem Statement II

From Fig. III-10, it can be seen that the following are designated as amplifier outputs: $\dot{x}_1$, $-x_1$, $x_1 - x_2$, $-\dot{x}_2$, $x_3 - x_2$, $\dot{x}_2 - \dot{x}_1$.

It has been established that we must restrict these outputs to lie in the range between $-$Reference and $+$Reference; thus, the scale factors must be selected. Choice of scale factors comes from a knowledge of (1) the maximum values of the problem variables (x_1, $\dot{x}_1$, x_2, $\dot{x}_2$, x_3) and (2) the value of computer reference (taken to be 1.0 unit).

While the maxima of x_1 and x_2 have been given, those of $\dot{x}_1$ and $\dot{x}_2$ must be estimated from the original equations. Finally, the difference terms can be estimated from the individual velocity and displacement maxima.

The original equations were

$$M_1 \ddot{x}_1 + D(\dot{x}_1 - \dot{x}_2) + K_1(x_1 - x_2) = 0$$

and

$$M_2 \ddot{x}_2 + D(\dot{x}_2 - \dot{x}_1) + K_1(x_2 - x_1) + K_2(x_2 - x_3) = 0$$

which can be written as

$$M_1 \ddot{x}_1 + D\dot{x}_1 + K_1 x_1 = D\dot{x}_2 + K_1 x_2$$
$$M_2 \ddot{x}_2 + D\dot{x}_2 + (K_1 + K_2)x_2 = D\dot{x}_1 + K_1 x_1 + K_2 x_3.$$

AMPLITUDE, TIME SCALING AND CHECKING PROCEDURES

Thus, the undamped natural frequency of the first equation is*

$$\omega_1 = \sqrt{K_1/M_1} = \sqrt{1000/25} \simeq 6.32$$

while the undamped natural frequency of the second equation is

$$\omega_2 = \sqrt{(K_1 + K_2)/M_2} \simeq 54.8.$$

Because damping tends to decrease the actual frequency, choose $\omega_1 = 5$ and $\omega_2 = 50$. Thus, $\dot{x}_1$ has a maximum of about 5 and $(\simeq \omega_1 x_{1\max})$ and $\dot{x}_2$ has a maximum of about 50. In order to assure accurate scaling, good bookkeeping practices should be adhered to. The following table demonstrates the bookkeeping required:

Problem Variable	Estimated Maximum	Computer Variable
x_1	1 ft.	$[x_1]$
$\dot{x}_1$	5 ft./sec.	$[0.2\dot{x}_1]$
x_2	1 ft.	$[x_2]$
$\dot{x}_2$	50 ft./sec.	$[0.02\dot{x}_2]$
$x_1 - x_2$	2 ft.	$[0.5(x_1 - x_2)]$
$\dot{x}_1 - \dot{x}_2$	50 ft.	$[0.02(\dot{x}_1 - \dot{x}_2)]$
$x_2 - x_3$	2 ft.	$[0.5(x_2 - x_3)]$

From this table, Eqs. (12) and (13) may be scaled appropriately by direct substitution as follows:

$$-\frac{d}{dt}\left[\frac{\dot{x}_1}{5}\right]5 = \left(+\frac{D}{M_1}\right)\left[\frac{\dot{x}_1 - \dot{x}_2}{50}\right]50 + \left(\frac{K_1}{M_1}\right)\left[\frac{x_1 - x_2}{2}\right]2 \quad (18)$$

$$+\frac{d}{dt}\left[\frac{\dot{x}_2}{50}\right]50 = \left(\frac{D}{M_2}\right)\left[\frac{\dot{x}_1 - \dot{x}_2}{50}\right]50 + \left(\frac{K_1}{M_1}\right)\left[\frac{x_1 - x_2}{2}\right]2$$

$$+ \left(\frac{K_2}{M_2}\right)\left[\frac{x_3 - x_2}{2}\right]2. \quad (19)$$

Equations (18) and (19) may now be rewritten as follows:

$$-\frac{d}{dt}\left[\frac{\dot{x}_1}{5}\right] = \left(\frac{10D}{M_1}\right)\left[\frac{\dot{x}_1 - \dot{x}_2}{50}\right] + \left(\frac{2K_1}{5M_1}\right)\left[\frac{x_1 - x_2}{2}\right] \quad (20)$$

$$\frac{d}{dt}\left[\frac{\dot{x}_2}{50}\right] = \left[\frac{D}{M_2}\right]\left[\frac{\dot{x}_1 - \dot{x}_2}{50}\right] + \left(\frac{K_1}{25M_2}\right)\left[\frac{x_1 - x_2}{2}\right]$$

$$+ \left(\frac{K_2}{25M_2}\right)\left[\frac{x_3 - x_2}{2}\right]. \quad (21)$$

---

*Techniques for evaluating natural frequencies and maximum values will be presented in Chapter V.

Equations (20) and (21) are the properly scaled equations; however, one must look at the integrator gains to be certain that they are between 0.1 and 10. If they are not between 0.1 and 10, then time scaling is necessary.

Examining the maximum value of the coefficients in parentheses, it is seen that they are all large:

$$8 \leq \frac{10D}{M_1} \leq 80 \qquad \frac{K_2}{25M_2} = 100 \qquad 10 \leq \frac{D}{M_2} \leq 100$$

$$\frac{2K_1}{5M_1} = 16 \qquad\qquad \frac{K_1}{25M_2} = 20.$$

This suggests that time scaling is required. The above equations can then be written as

$$-\frac{d}{d\tau}\left[\frac{\dot{x}_1}{5}\right] = 10\left(\frac{D}{M_1\beta}\right)\left[\frac{\dot{x}_1 - \dot{x}_2}{50}\right] + 10\left(\frac{2K_1}{50M_1\beta}\right)\left[\frac{x_1 - x_2}{2}\right] \quad (22)$$

$$\frac{d}{d\tau}\left[\frac{\dot{x}_2}{50}\right] = 10\left(\frac{D}{10M_1\beta}\right)\left[\frac{\dot{x}_1 - \dot{x}_2}{50}\right] + 10\left(\frac{K_1}{250M_2\beta}\right)\left[\frac{x_1 - x_2}{2}\right]$$

$$+ 10\left(\frac{K_2}{250M_2\beta}\right)\left[\frac{x_3 - x_2}{2}\right]. \quad (23)$$

A value of $\beta = 10$ will make most of these gains "reasonable," i.e., between 0.1 and 10. The final scaled computer diagram is shown in Fig. IV-9. Note the addition of potentiometers between integrators and into summers to make the scaling complete.

Problem Statement III

In the process of programming, scaling and patching a problem, numerous opportunities for error arise. Since even one error in a program can invalidate the entire result, it is absolutely necessary to detect and correct these errors before actual computation starts. The method of detecting and correcting programs is called Static Check.

Perform a Static Check for the Automobile Suspension problem where:

$$x_1 = +1.0 \text{ ft.}$$
$$\dot{x}_1 = +4.0 \text{ ft./sec.}$$
$$x_2 = +0.8 \text{ ft.}$$
$$\dot{x}_2 = +40 \text{ ft./sec.}$$

and parameter values are as in Part 1 with $D = 200$ lb./ft./sec.

After the Static Check is completed, patch the problem and investigate the effect of changing the shock absorber damping coefficient, D.

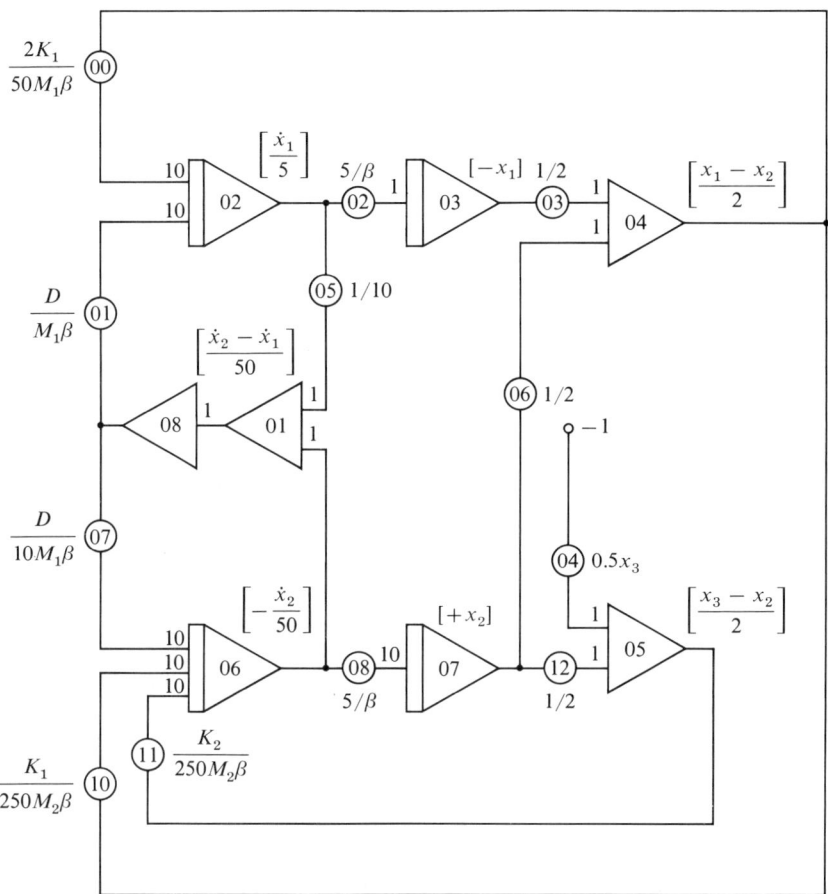

Fig. IV-9. Completely Scaled Auto Suspension System ($\beta = 10$)

SOLUTION TO PROBLEM STATEMENT III

The performance of a complete Static Check involves two parts: a *program check* and a *circuit check*. The solution presented here will depict the program check. The circuit check involves patching the program on the computer and measuring voltages to compare to calculations.

For a complete *program check,* calculations on the computer circuit itself must be made in addition to calculations from the original unscaled equations.

a. Calculations based on the program can be performed directly on a copy of the circuit diagram if the outputs of all integrators and all pot settings are known. The integrator outputs are:

$$[\dot{x}_1/5],\ -[x_1],\ -[\dot{x}_2/50],\ [x_2]$$

and their values are found from the values assumed for x_1, $\dot{x}_1$, x_2 and $\dot{x}_2$ in the statement of the problem; pot settings are calculated from the assumed values of the parameters (see problem statement). Figure IV-10 shows the diagram with these quantities marked in appropriate places.

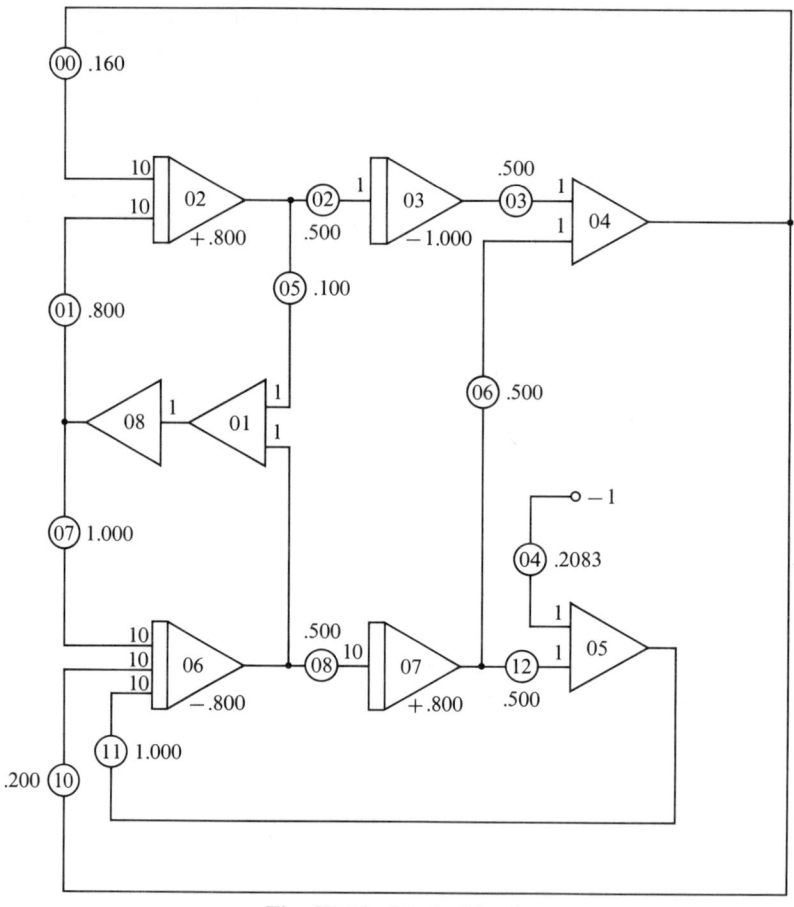

Fig. IV-10. Static Check

Now without any other information except that which is shown in Fig. IV-4, all amplifier outputs can be calculated. Also, all derivative inputs of integrators can be found. For example,

$$A04 = -\{(1/2)(-1) + (1/2)(+.8)\} = 0.100$$
$$A05 = -\{(1/2)(+.8) + (.2083)(-1)\} = -0.1917$$

$$A01 = -\{(1/10)(.8) + (-.8)\} = +0.720$$
$$A08 = -A01 = -0.720$$

and

$$D03 = (A02)(P02)(1) = (.8)(.500)(1) = 0.400$$
$$D07 = 10(.500)(-.8) = -4.000$$
$$D02 = 10(.160)(.100) + 10(.800)(-.720) = -5.600$$
$$D06 = 10(1.000)(-.720) + 10(.200)(.100) + 10(1.000)(-.1917)$$
$$= -8.917.$$

b. Calculations based on the original problem are performed for amplifier outputs and derivative values as follows: the output of each amplifier is known in terms of the problem variables and a scale factor. Knowing the assumed values of these allows the calculation of these outputs without referring to the diagram at all. The scaling table lists the computer variables (i.e., the amplifier outputs) as

$$[x_1]$$
$$[\dot{x}_1/5]$$
$$[x_2]$$
$$[\dot{x}_2/50]$$
$$[(x_1 - x_2)/2]$$
$$[(\dot{x}_1 - \dot{x}_2)/50]$$
$$[(x_2 - x_3)/2]$$

and values for $x_1, x_2, x_3, \dot{x}_1, \dot{x}_2$ are known from the problem statement. Thus,

$$A01 = [(\dot{x}_2 - \dot{x}_1)/50] = (40 - 4)/50 = 0.720$$
$$A02 = [\dot{x}_1/5] = 4/5 = 0.800$$
$$A03 = [x_1] = -1.000$$
$$A04 = [(x_1 - x_2)/2] = (1 - .8)/2 = 0.100$$
$$A05 = [(x_3 - x_2)/2] = (5/12 - 0.8)/2 = -0.1917$$
$$A06 = -[\dot{x}_2/50] = -40/50 = -0.800$$
$$A07 = [x_2] = +0.800$$
$$A08 = [(\dot{x}_1 - \dot{x}_2)/50] = (4 - 40)/50 = -0.720$$
$$D03 = -\frac{d}{d\tau}[-x_1] = +\frac{\dot{x}_1}{\beta} = 4/10 = 0.400$$
$$D07 = -\frac{d}{d\tau}[+x_2] = -\dot{x}_2/\beta = -40/10 = -4.000$$
$$D02 = -d[\dot{x}_1/5]/d\tau = -\ddot{x}_1/5\beta$$

$$= (1/5\beta)\{(D/M_1)(\dot{x}_1 - \dot{x}_2) + (K_1/M_1)(x_1 - x_2)\}$$
$$= (1/50)\{(200/25)(4 - 40) + (1000/25)(1.0 - 0.8)\}$$
$$= -5.600$$

and

$$D06 = -d[-\dot{x}_2/50]/d\tau = +\ddot{x}_2/50\beta$$
$$= (1/50\beta)\{(D/M_2)(\dot{x}_1 - \dot{x}_2) + (K_1/M_2)(x_1 - x_2)$$
$$+ (K_2/M_2)(x_3 - x_1)\}$$
$$= (1/500)\{(200/s)(4 - 40) + (1000/2)(1.0 - 0.8)$$
$$+ (5000/2)(0.4167 - 0.8)\}$$
$$= -8.917.$$

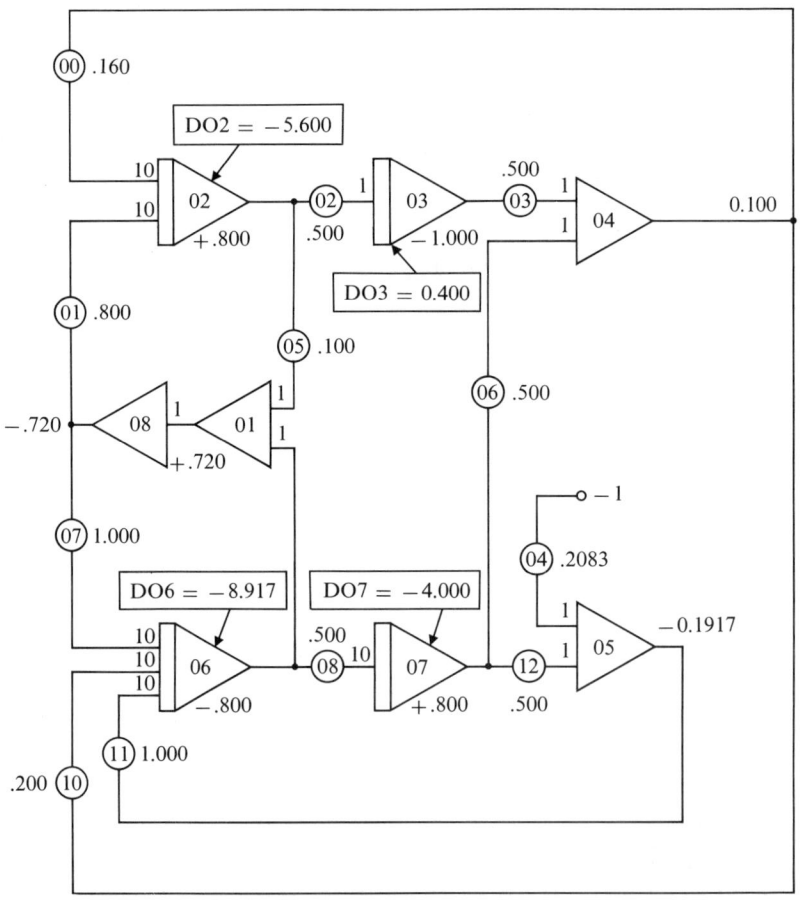

Fig. IV-11

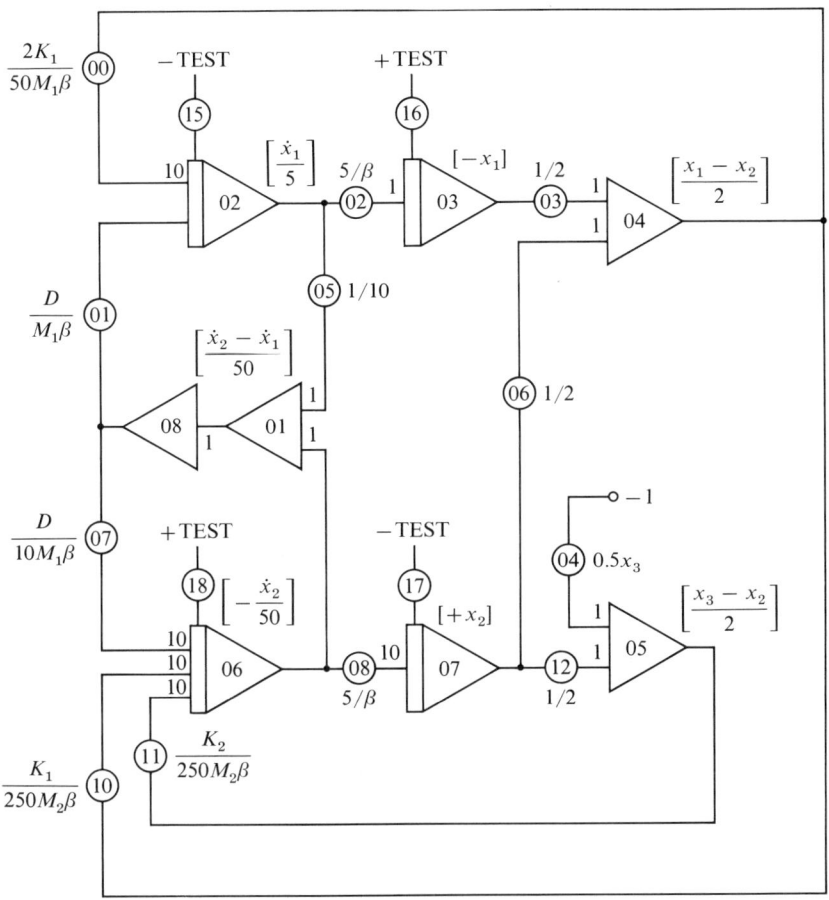

Fig. IV-12. Final Computer Program

c. Comparisons are now in order. Checking the values calculated in (b) with the ones calculated on the diagram in (a) shows 100% agreement. Thus, it is safe to say that the program of Fig. IV-3 truly represents the problem described in Part I. These values may now be entered on pot and amplifier sheets.

d. The final effort in checking involves patching the program, setting pots, establishing the selected integrator outputs, and measuring all amplifier outputs and derivatives. The measured results should agree with calculated values to better than 1% or an error is indicated. Proper integrator outputs for checking this problem are assured by the addition of test IC's for the four integrators through pots 15, 16, 17, and 18 as shown in Fig. IV-12.

These values are shown in Fig. IV-11, the completed program check.

POTENTIOMETER ASSIGNMENT SHEET

Date _____   Problem __Auto Suspension__

Pot. No.	Parameter Description	Setting Static Check	Static Check Output Voltage	Setting Run Number 1	Notes	Pot. No.
00	$2K_1/50M_1\beta$	0.160		0.160		
01	$D/M_1\beta$	0.800		0.400		
02	$5/\beta$	0.500		0.500		
03	$1/2$	0.500		0.500		
04	$x_3/2$	0.2083		0.2083		
05	$1/10$	0.100		0.100		
06	$1/2$	0.500		0.500		
07	$D/10M_2\beta$	1.000		0.500		
08	$5/\beta$	0.500		0.500		
09						
10	$K_1/250M_2\beta$	0.200		0.200		
11	$K_1/250M_2\beta$	1.000		1.000		
12	$1/2$	0.500		0.500		
13						
14						
15	TEST $\dot{x}_{10}/5$	0.800		0		
16	TEST x_{10}	1.000		0		
17	TEST x_{20}	.800		0		
18	TEST $\dot{x}_{20}/50$	.800		0		
19						
20						
21						
22						
23						
24						

AMPLIFIER ASSIGNMENT SHEET

Date_____   Problem __Auto Suspension__

Amp. No.	FB	Output Variable	Static Check				Notes
			Calculated		Measured		
			Check Pt.	Output	Check Pt.	Output	
00	Σ	$[(\dot{x}_2 - \dot{x}_1)/50]$		0.720			
01	$\int$	$[x_1/5]$	+0.560*	+0.800			
02	$\int$	$[-x_1]$	−0.400	−1.000			
03	Σ	$[(x_1 - x_2)/2]$		0.100			
04	Σ	$[(x_3 - x_4)/2]$		−0.192			
05	$\int$	$[-\dot{x}_2/50]$	+0.892*	−0.800			
06	$\int$	$[+x_2]$	+0.400*	+0.800			
07	−1	$[(x_1 - x_2)/50]$		−0.720			
08							
09			Output of Check Amplifier				
10							
11							
12							*Check Amplifier Gain = −1/10
13							
14							
15							
16							
17							
18							
19							
20							

CONCLUSION

This chapter concerned itself with the practical aspects of amplitude and time scaling. After the generalizations of amplitude and time scaling were considered, a specific problem was discussed. Although amplitude scaling was examined, no technique for examining maximum values was considered. This discussion will be reserved for Chapter V.

CHAPTER V

Linear Systems Analysis

AN ANALOG COMPUTER or an Analog/Logic Computer are very good devices for simulating both linear and non-linear systems. Since there is actually no such thing as a linear system, computers usually simulate non-linear systems or systems which are made to appear linear. This chapter will be concerned with the definition of linear and non-linear systems in addition to the practical considerations of programming and estimating some maximum values. Only those aspects of linear systems that are directly applicable to Analog/Logic Computers will be considered. No attempt will be made to exhaust the entire subject of linear systems.

Linear and Non-Linear Systems

A system is linear if it satisfies the principle of superposition. If one thinks of a system as depicted in Fig. V-1, where $X1$ and $X2$ are the input variables, then it can be said that a system is *linear* if and only if

$$H(\alpha X1 + \beta X2) = \alpha HX1 + \beta HX2 \tag{1}$$

Fig. V-1. Physical System

where α and β are any constants, and $X1$ and $X2$ are any input signals.* It is important to remember the assumption that $X1$ and $X2$ are applied at $t = -\infty$ and that the system is unexcited at that time. The foregoing definition, with $\alpha = \beta = 1$, becomes:

$$H(X1 + X2) = HX1 + HX2 \tag{2}$$

which is called the property of additivity, and asserts that the response to a sum of two inputs is equal to the sum of the two responses. With $X2 \equiv 0$ the definition becomes

$$H(\alpha X1) = \alpha HX1 \tag{3}$$

which is called the property of homogeneity and asserts that the response to a constant multiple of any input is equal to the response to the input multiplied by the same constant. Thus a linear system possesses both the property of *additivity* and the property of *homogeneity*. It is easy to show that any system which possesses these properties is linear. Hence, a system is linear if and only if it is both additive and homogeneous.

By induction it can be demonstrated that in a linear system

$$H\left(\sum_{i=1}^{n} a_i x_i\right) = \sum_{i=1}^{n} \alpha i H x_i \tag{4}$$

provided that the upper index n is finite.

Linear systems may be classified into lumped and distributed systems. They may also be classified as time-invariant and time-variant systems.

A system is a collection of individual elements interconnected in a particular way. A lumped system consists of lumped elements. In a lumped model the energy in the system is considered to be stored or dissipated in distinct isolated elements (resistors, capacitors, inductors, masses, springs, dashpots, etc.). Also, it is assumed that the disturbance initiated at any point is propagated instantaneously at every point in the system. In contrast to lumped systems there are distributed systems such as transmission lines, waveguides, antennas, semiconductor devices, pipe lines, neuron systems, etc. In such systems it takes a finite amount of time for a disturbance at one point to be propagated to another point. Thus, one has to deal not only with the independent variable time (t) but also the space variable x. The descriptive equations for distributed systems are therefore partial differential equations in contrast to ordinary differential equations describing lumped systems.

Whether a system is lumped parameter or distributed parameter it must be described by differential equations which may be obtained in any one or combination of the following ways:

---

* For a more precise definition, see L. A. Zadeh, An Extended Definition of Linearity, Proc. IRE, Vol. 49, p. 1452, Sept., 1968 and Vol. 50, p. 200, Feb., 1962.

(1) Equilibrium methods
    Σ FORCES = 0
    Σ TORQUES = 0
(2) Hamiltonian method
    potential and kinetic energies of the system are written in terms of generalized set of *position* and momentum variables.
(3) Lagrangian techniques
    difference between potential and kinetic energy is put into Euler-Lagrange equation.
(4) Kirchoff's laws
    application of *loop* and *nodal* analysis to electrical, mechanical, biological, hydraulic, etc. systems.
(5) System theory techniques
    using Laplace, Fourier and z-transforms in addition to system flow diagram techniques one can develop system equations.

As already mentioned, linear systems can also be classified into time-invariant and time-variant systems. The systems whose parameters do not change with time are called constant-parameter or time-invariant systems. Most of the systems observed in practice belong to this category. Linear time-invariant systems are characterized by linear equations (algebraic, differential, or difference equations) with constant coefficients. Electrical circuits using passive elements are an example of time-invariant systems. On the other hand, we have systems whose parameters change with time and are therefore called variable parameter or time-invariant (also time-dependent) systems. Linear time-variant systems are characterized by linear equations with time-dependent coefficients in general. An example of a simple linear time-variant system is shown in Fig. V-2.

The equation below, describing this system, is derived from Kirchoff's loop analysis

$$L\frac{di(t)}{dt} + R(t)i(t) = f(t) \tag{5}$$

where $f(t)$ is the voltage driving function.

This is a linear, time-variant differential equation because it obeys the superposition principle, but one of the terms in the differential equation is time dependent.

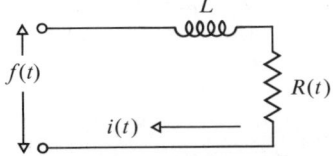

Fig. V-2. Linear Time-Variant System

An example of a linear constant coefficient differential equation is indicated in Eq. (6).

$$\frac{d^2\gamma}{dt^2} + a\frac{d\gamma}{dt} + b\gamma = f(t). \tag{6}$$

All coefficients of this differential equation are constant whereas those in Eq. (5) were not.

Generally, a constant coefficient linear differential equation is described as in Eq. (7).

$$a_n\left(\frac{dx^n}{dt^n}\right)^m + a_{n-1}\frac{dx^{n-1}}{dt^{n-1}} + \cdots a_o x = f(t). \tag{7}$$

In this general equation, $a_o, a_1, \ldots a_n$ are constants, the order of the equation is n and the degree $m = 1$. If m is not equal to 1 ($m \neq 1$) this becomes a non-linear differential equation.

There is really no such thing as an ordinary differential equation describing a system. That is, systems found in nature are generally described by partial differential equations. Indicated in Eq. (8) is the general form for an mth order partial differential equation. Although this is the form of the equation describing most physical systems the analog computer solution of this equation will not be considered in this introductory material.

$$\frac{\partial m_x}{\partial t^m} + a_{m-1}\frac{\partial m - 1_x}{\partial t^{m-1}} + \cdots + a_o x = \frac{\partial n_x}{\partial u^n}. \tag{8}$$

Estimating Maximum Values

This chapter will not attempt to present linear systems analysis techniques. This material may be found in many of the references at the end of the chapter in addition to introductory electrical engineering texts.

It was mentioned that Eq. (7) depicted the general form of an nth order linear differential equation with constant coefficients. Special cases of this general differential equation will now be considered and then by induction, a general technique will be developed for analyzing the nth order equation.

Consider once again, Eq. (7) rewritten here

$$a_n\frac{dx^n}{dt^n} + a_{n-1}\frac{dx^n}{dt^{n-1}} + \cdots a_o x = f(t).$$

For the spring, mass, damper system the following order differential equation can be written (refer to Chapter IV Eq. [1]

$$a\ddot{x} + b\dot{x} + cx = 0 \tag{9}$$

LINEAR SYSTEMS ANALYSIS

where: $f(t) = 0$ (forcing function)
  a — mass of system
  b — dashpot coefficient
  c — spring coefficient.

Equation (9) is of the form of Eq. (7) for $n = 2$.

Equation (9) can be written in two other forms. For $b = 0$, Eq. (9) becomes

$$a\ddot{x} + c\dot{x} = 0 \qquad (10)$$

and for $a = 0$, Eq. (9) becomes

$$b\dot{x} + cx = 0. \qquad (11)$$

Equations (9), (10) and (11) are special cases of Eq. (7).

If one considers Eq. (11), a solution may be obtained by the separation of variables, that is

$$\frac{b\,dx}{dt} = -cx \qquad (12)$$

$$\frac{b\,dx}{x} = -c\,dt \qquad (13)$$

$$b \ln x = -ct \qquad (14)$$

$$\ln x = -\left(\frac{c}{b}\right)t \qquad (15)$$

$$x(t) = e^{-(c/b)t}. \qquad (16)$$

From Eq. (16) at

$$t = 0$$
$$x(o) = 1$$

then, the curve depicting the solution to this equation is shown in Fig. V-3.

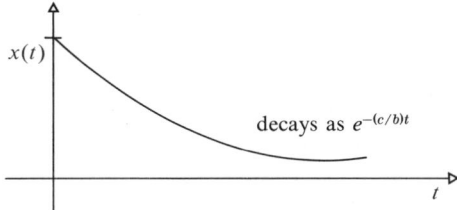

Fig. V-3. Solution to Equation 11

Generally, the solution to Eq. (11) is

$$x(t) = x(o)e^{-(c/b)t} \qquad (17)$$

where the maximum value is $x(o)$ and the maximum values of the derivatives can be found by taking successive derivatives of (17).

Let us assume that Eq. (10) has the following solution, that is

$$\ddot{x} + \frac{e}{a}x = 0 \qquad (18)$$

$$x(t) = x(o) \sin \sqrt{\frac{c}{a}} t \,. \qquad (19)$$

If this equation is substituted back into the original equation it can be shown to satisfy (18).

The question that a programmer is interested in at this point is: What is the maximum value of the outputs of the amplifiers? Once this question has been resolved for this homogeneous ($f(t) = 0$) differential equation a general technique can be developed for finding the maximum value of any nth order differential equation.

Making the following substitution:

$$\omega_n = \sqrt{\frac{c}{a}} \quad \text{natural frequency}$$

then Eq. (18) becomes:

$$\ddot{x} + \omega_n^2 x = 0\,. \qquad (20)$$

The analog computer circuit for the solution to this equation is shown in Fig. V-4.

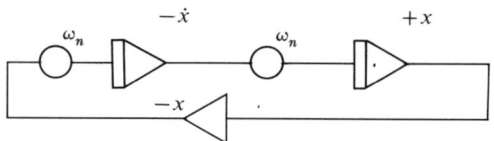

Fig. V-4. Analog Computer Circuit for the Solution of Equation 20

Now, one must determine the maximum values of the outputs of amplifiers, namely $\dot{x}$ and x.

Since $\sin \omega_n t$ varies between ± 1, the maximum value of $x(t)$ is

$$|x(t)| = x(o) \qquad (21)$$

from

$$x(t) = x(o) \sin \omega_n t \qquad (22)$$

$$\frac{dx(t)}{dt} = \dot{x} = x(o)\omega_n \cos \omega_n t\,.$$

Therefore, the maximum value of $\dot{x}$ is

$$|\dot{x}| = x(o)\omega_n\,. \qquad (23)$$

Since the nth derivative of
$$x(t) = x(o) \sin \omega_n t$$
is
$$\frac{dx^n(t)}{dt^n} = x(o)\omega_n^n \sin \omega_n t \text{ for } n - \text{even} \tag{24}$$
and
$$\frac{dx^n(t)}{dt^n} = x(o)\omega_n^n \cos \omega_n t \text{ for } n - \text{odd} \tag{25}$$

it can be concluded that the maximum values of all derivatives can be obtained for a general nth order derivative by simply solving for the natural frequency

$$\omega_n = \sqrt[n]{\frac{a_o}{a_n}}$$

and developing the following progression:
$$|x(t)| = x(o)$$
$$|\dot{x}(t)| = \omega_n \, x(o)$$
$$|\ddot{x}(t)| = \omega_n^2 x(o)$$
$$\vdots \quad \vdots \quad \vdots$$
$$|x^n(t)| = \omega_n^n x(o).$$

Therefore, to estimate the maximum values of the general homogeneous differential equation

$$a_n \frac{dx^n}{dt^n} + a_{n-1} \frac{dx^{n-1}}{dt^{n-1}} + \cdots a_o x = 0$$

the *general* rules are:

(1) calculate
$$\omega_n = \sqrt[n]{\frac{a_o}{a_n}}$$

(2) estimate the maximum values of the higher order derivatives by using the rule
$$|\dot{x} \text{ max}| \approx \omega_n x(o)$$
$$|\ddot{x} \text{ max}| \approx \omega_n^2 x(o)$$
$$|x^n \text{ max}| \approx \omega_n^n x(o) \, .$$

This method is quite easy to use and results in approximately equal distribution of gains around the main computing loop. The reader is referred to the example of estimating maximum values indicated in the automobile suspension system problem of Chapter IV.

Since every problem is unique, the estimation of maximum values is *unique* for every problem and must be handled as such. The numerous problems in the Appendix indicate other methods used for estimating maximum values.

The preceding analysis was for unforced systems $f(t) = 0$; suppose, however, the forcing function was not zero but one of the following.

(1) Step Input
(2) Sinewave Input
(3) Arbitrary Input

The question is; how can one estimate maximum values in this case? The next section will be devoted to this problem.

FORCED SYSTEMS ($f(t) \neq 0$)

This section discusses the equation with a forcing function

$$a_n \frac{dx^n}{dt^n} + a_{n-1} \frac{dx^{n-1}}{dt^{n-1}} + \cdots a_o x = f(t)$$

where $f(t) \neq 0$.

Relation between Forced and Unforced Systems

One of the first things to be learned about solving such forced-system equations is that it is a very good idea to be on familiar terms with the corresponding *unforced* equation, for the two are very closely related.

For example, suppose the unforced equation has already been solved completely, with all the time-constants and natural frequencies characterizing the system found and, hence, all solutions obtained. Then, if *just one* solution for the forced equation can be found, then *all* solutions for a given forcing function can be found.

To see this, suppose there are two functions, $x_1(t)$ and $x_2(t)$, which satisfy the same linear, forced equation with the same forcing function, $f(t)$. This means

$$a_n \frac{d^n x_1}{dt^n} + a_{n-1} \frac{d^{n-1} x_1}{dt^{n-1}} + \cdots + a_1 \frac{dx_1}{dt} + a_o x_1 = f(t) \quad (26)$$

and

$$a_n \frac{d^n x_2}{dt^n} + a_{n-1} \frac{d^{n-1} x_2}{dt^{n-1}} + \cdots + a_1 \frac{dx_2}{dt} + a_o x_2 = f(t). \quad (27)$$

Subtracting, we discover that

$$a_n \frac{d^n(x_1 - x_2)}{dt^n} + a_{n-1} \frac{d^{n-1}(x_1 - x_2)}{dt^{n-1}} + \cdots$$

$$+ a_1 \frac{d(x_1 - x_2)}{dt} + a_o(x_1 - x_2) = 0. \quad (28)$$

In other words, *the difference between two solutions of the forced equation satisfies the unforced equation.*

This general principle means that the *general* solution of the forced system can be obtained by finding just *one* solution (usually called a *particular integral*) and adding to it the general solution of the unforced system (usually called the *complementary function*) which we already know how to obtain.

The problem of finding the general solution in the forced case, therefore, boils down to the problem of finding any solution. Trial-and-error is often very successful here, and there are general methods available to use when trial-and-error fails.

As a starter, consider several cases where $f(t)$ has a particularly simple form.

SOLUTION BY STEP INPUT

Suppose $f(t)$ is a step input of height, k. Then we have to find a solution to the equation

$$a_n \frac{d^n x}{dt^n} + a_{n-1} \frac{d^{n-1}}{dt^{n-1}} + \cdots + a_1 \frac{dx}{dt} + a_o x = k. \qquad (29)$$

One idea immediately suggests itself: a constant value for x. It is not hard to see that if $x = k/a_o$, then the equation is satisfied. This is a trivial solution and, in a given physical situation, it almost certainly will *not* be the solution of interest, but no matter! All solutions can be obtained by adding to the constant, k/a_o, the general solution of the unforced equation. The result will contain enough arbitrary constants to match initial conditions for x and its first $n - 1$ derivatives, and thus find the required solution.

Typical step responses for the first- and second-order systems are given in Figs. V-5 and V-6. Note that overshoot and oscillations occur only in the underdamped second-order case. The overshoot, if the system is stable, is never more than 100%, and even this value is only approached in very lightly damped second-order systems. This fact is often useful in estimating maximum values for scaling purposes. The technique usu-

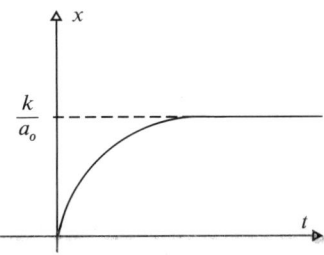

Fig. V-5. Solution to the First-Order Equation $a\dot{x} + bx = k$

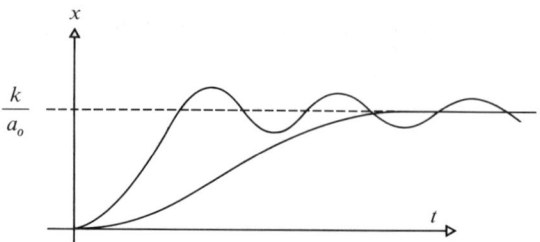

Fig. V-6. Solution to the Second-Order Equation $a\ddot{x} + b\dot{x} + cx = k$

ally used for estimating maximum values is known as the equal coefficient rule.

If Eq. (29) is considered for the simple case
$$f(t) = k = Au_{-1}(t) \tag{30}$$
where
$$Au_{-1}(t) = \begin{cases} 0 & t \leq 0 \\ A & t \geq 0 \end{cases}$$
and zero initial conditions, the maximum value of the nth derivative often occurs at $t = 0$ and is equal to
$$X^{(n)}m = \frac{A}{a_n}. \tag{31}$$

The final value of $x(t)$ is A/a_0 if the system is stable and if no roots of the characteristic equation occur on the imaginary axis. However, the maximum value of $x(t)$ must be taken as $2A/a_0$ in case roots do occur on the imaginary axis.

With normalized variable
$$A \frac{d}{dt^n}\left[\frac{x^n}{A/a_n}\right] + x_m^{n-1} a_{n-1} \frac{d^{n-1}}{dt^{n-1}}\left[\frac{x}{x_m^{n-1}}\right] + \cdots$$
$$+ 2A\left[\frac{x}{2A/a_0}\right] = Au_{-1}(t) \tag{32}$$

the maximum values may be estimated by using values that cause all the coefficients in the normalized equation to be approximately equal, with the restriction that the resultant assumed maxima constitute a monotonic set; that is, that the values continually increase (or decrease) for each derivative in turn. If the initial conditions are all zero and if $f(t)$ is a step function, the coefficient of the (x/x_m) term should be twice that of the others.

The maximum values must usually be modified if initial conditions are not zero, and in this case the coefficient of $f(t)$ will be less than unity.

As an example, consider the second order differential equation

$$a\ddot{x} + b\dot{x} + cx = Au_{-1}(t) \tag{33}$$

then

$$\ddot{x} + \left(\frac{b}{a}\right)\dot{x} + \left(\frac{c}{a}\right)x = Au_{-1}(t) \tag{34}$$

where

$$\dot{x}(o) = x(0) = 0.$$

The normalized equation is

$$\ddot{x}_m\left(\frac{\ddot{x}}{\ddot{x}_m}\right) + \left(\frac{b}{a}\right)\dot{x}_m\left(\frac{\dot{x}}{\dot{x}_m}\right) + \left(\frac{c}{a}\right)x_m\left(\frac{x}{x_m}\right) = Au_{-1}(t) \tag{35}$$

where the subscript m designates maximum value.

Then using the rule of equal coefficients one would choose

$$\ddot{x}_m = A; \quad \dot{x}_m = \frac{Aa}{b}; \quad x_m = \frac{2Aa}{c} \tag{36}$$

therefore

$$\dot{x}_m = \frac{x_m c}{2b} \tag{37}$$

$$\ddot{x}_m = \frac{x_m c}{2a}. \tag{38}$$

Since x_m is known, it can be seen that the maximum values of the higher order derivatives are determined by x_m and the constant coefficients in the equation.

If one is examining the behavior of a system to a step input the equal coefficient rule should be utilized.

Sinewave Input

Suppose the input is a sinewave, $f(t) = \sin \omega t$. Then, both experience and intuition (mathematical or physical) will lead most people to expect a sinusoidal output.

Usually, this turns out to be the case. One may try for a sinusoidal solution by assuming $x(t) = A \sin(\omega t + \emptyset)$ and then differentiating and substituting to find the amplitude, A, and the phase angle, $\emptyset$, of the output relative to the input. With luck, two equations in the two unknowns will result which can be solved for A and $\emptyset$. The algebra gets a bit complicated but it can be simplified greatly by regarding the input as the imaginary part of the complex exponential, $e^{j\omega t} = \cos \omega t + j \sin \omega t$. For the purposes of this volume, it will suffice to observe that a sinusoidal input into a linear, constant co-efficient system generally produces a sinewave output, and that the amplitude and phase of the output relative to the input can be calculated algebraically.

There is one case in which the above method for finding a particular integral does *not* produce a pure sinewave. This is the case where the unforced system has a normal mode consisting of a pure (i.e., undamped) sinewave with the same frequency as the input. In this case, the solution contains terms like $(A + Bt)(\sin \omega t)$ indicating oscillations whose amplitude increases without bound. In case the normal mode is slightly damped, we have oscillations with a finite, but very large amplitude. In either case, the system is said to be in *resonance*.

Physically, resonance means that if a system is driven by an input whose frequency is close to a natural frequency of the system (that is, a frequency at which the system would tend to oscillate even if unforced), the result will be an output of very large amplitude. For example, in pushing a child in a swing, large amplitude oscillations are necessary to satisfy the child. To achieve these with minimum effort, the pusher soon learns to time his pushes so that each push reinforces the motion of the swing rather than opposing it. This means that the pusher must match the frequency of his pushes to the natural frequency of the swing.

Resonance is also important in structural design. If an airplane happens to have lightly-damped normal modes whose frequency is close to the rpm of the engines, the resulting vibrations can build up in amplitude until they tear the plane apart. It is the responsibility of the designer to see that this resonance phenomenon does not happen. The same sort of thing can occur, say, when a detachment of infantry march in cadence across a bridge. Often, the leader will tell his men to break step and just *walk* across to prevent this sort of catastrophe.

Resonance is also important in tuning a radio. To get the maximum output for a given input, it is desirable to adjust the tuner section of the radio so that its natural frequency is as close as possible to the frequency at which the station is actually broadcasting.

Arbitrary Input

Suppose $f(t)$ is an arbitrary function given by some complicated analytical expression or by a graph or table of values. To find one solution (a particular integral), in this case, there exists a number of techniques, some of which are listed below. For more details, see Reference 1.

(1) The convolution integral: This technique gives the solution in terms of an integral which may or may not be easy to evaluate. It has the advantage of working with tabular or graphical inputs, in which case the integral must be evaluated by some numerical method (Simpson's Rule, counting squares, etc.).

(2) Harmonic analysis: If the input is periodic but not sinusoidal, it may be broken down into an infinite sum of sinewaves. Using the principle of superposition, each sinewave may be handled separately and then added to obtain the resultant output.

(3) <u>Integrating Factors</u>: This method works nicely with first-order systems, including those with variable co-efficients.

(4) <u>Undetermined coefficients</u>: It has already been shown how this works in tackling the unforced equation by assuming an exponential solution, e^{kt}, and solving for k. Sometimes it works very well for the forced case also.

(5) <u>Variation of parameters</u>: This method is similar to the method of undetermined coefficients. It has the advantage of working in the case of time-varying coefficients, and the disadvantage of requiring the solution of the corresponding *unforced* case first.

(6) <u>Laplace Transforms</u>: This method is very handy for functions that have Laplace transforms that can be easily managed. See Reference 3 for details.

(7) <u>Infinite series</u>

(8) <u>Guessing</u>: This is the easiest method to apply, if it works. If it does, more complicated techniques are not necessary.

LINEAR SYSTEMS WITH TIME-VARYING COEFFICIENTS

Suppose the co-efficients, a_n, in Eq. (7) are not constants but known functions of time. Then, the exponentials and sinewaves encountered in the constant-coefficient case will not always work. Other methods, such as infinite series, integral transforms and integrating factors must be used to find particular solutions. However, it is worth pointing out that the principles of superposition and proportionality still work. If enough different solutions of the unforced equation can be found, they can be multiplied by arbitrary constants and added to obtain the general solution. If the unforced system can be solved completely and just *one* solution of the forced equation can be found, then the general solution for that forcing function can be obtained by adding the complementary function (general solution of the unforced system) to the particular integral (solution of the forced system), just as was done in the constant-coefficient case. A more detailed discussion of time-varying systems is beyond the scope of this volume.

Conclusion

This chapter concerned itself with an introduction to linear system analysis and techniques for evaluating the maximum values of differential equations. In order for one to get a completely accurate picture of the system to be analyzed the techniques considered in this chapter must be followed. No attempt was made to present all of the techniques. This is left for other books to dwell upon.

References

1. R. P. Agnew, *Differential Equations* (New York: McGraw-Hill Book Company, 1960).
2. S. Fifer, *Analog Computation* (New York: McGraw-Hill Book Company, 1961).
3. B. P. Lathi, *Signals, Systems and Communication* (New York: John Wiley, 1965).
4. Friedland Schwarz, *Linear Systems* (New York: McGraw-Hill Book Company, 1965).
5. H. T. Milhorn, *The Application of Control Theory to Physiological Systems* (Philadelphia: Saunders, 1966).
6. C. B. Kuo, *Automatic Control Systems* (Englewood, N.J.: Prentice Hall, Second Edition, 1967).
7. G. A. Bekey, and W. J. Karplus, *Hybrid Computation* (New York: John Wiley, 1968).

CHAPTER VI

Non-Linear Programming Elements and Circuits

IT WAS POINTED OUT quite explicitly in Chapter II that analog computation is concerned with solving the nth order differential equation depicted below:

$$a_n\left(\frac{dx^n}{dt^n}\right)^m + a_{n-1}\frac{dx^{n-1}}{dt^{n-1}} + \cdots a_o x = f(t). \tag{1}$$

If all of the a_n's are constants and all of the derivatives have a degree $m = 1$, the equations are linear. However, if $m \neq 1$ or the a_n coefficients are not constants, then the equation is considered to be non-linear. The mathematical operations performed by non-linear components are:

(1) multiplication and division of variables,
(2) the generation of arbitrary functions, and
(3) the mechanization of constraints and elementary logic operations.

A consideration of components that perform the operations indicated above will now be undertaken.

Multiplication and Division of Variables

The question may arise as to how one can perform the multiplication of two variables x and y to get a function:

$$f(x,y) = xy. \tag{2}$$

To do this, one needs an electronic device which will perform the following function:

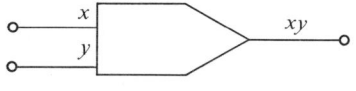

Fig. VI-1

Not until we discuss the generation of arbitrary functions will we be concerned with what is inside of the box.

If one had an equation to solve which was of the following form:

$$\frac{d^2x}{dt^2} + \frac{dx}{dt} + xt = c \qquad (3)$$

a solution could very readily take place using Fig. VI-2 and the following formulation:

$$\frac{dx^2}{dt^2} = -\frac{dx}{dt} - xt + c \cdot \qquad (4)$$

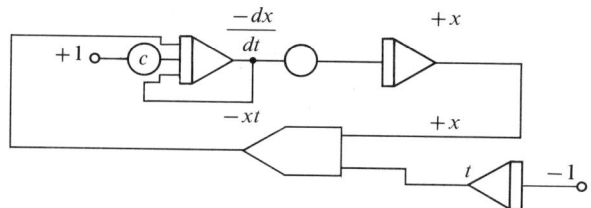

Fig. VI-2. Solution to Equation 3

Equation (3) is a non-linear differential equation which can only be solved using the component depicted in Fig. VI-1. Appendix E illustrates the use of a multiplier in solving a d.c. Servo Simulation. One may program using only a block diagram and never knowing the internal workings of the multiplier. In order to program from this point of view it should be noted that *no scaling is necessary* with a multiplier. If the input variables are scaled, then the output of the multiplier will be scaled also.

Multiplying devices may also be used to perform other non-linear operations if utilized properly. Figure VI-3 illustrates how a multiplying or squaring device may be used to perform division.

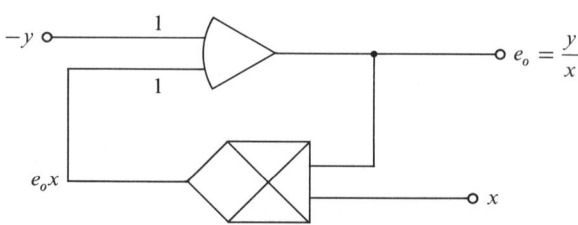

Fig. VI-3. Division Circuit

The operation of this circuit can be explained by considering the fact that:

$$\Sigma(\text{input volt})(\text{gain}) = 0 \tag{5}$$

in a high gain amplifier. Then

$$-y + e_o x = 0 \tag{6}$$

and

$$e_o = \frac{y}{x}. \tag{7}$$

If $x \neq 0$, this circuit will work very well.

Another circuit utilizing the squaring device is the square root circuit. Figure VI-4 shows this circuit

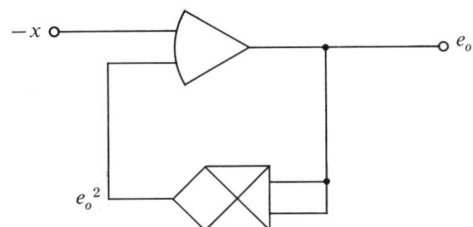

Fig. VI-4. Square Root Circuit

where

$$e_o^2 - x = 0 \tag{8}$$

thus

$$e_o = \sqrt{x} \tag{9}$$

where x is not imaginary or negative.

Generally, it can be said that the inverse of a function may be obtained by placing the non-linear component in the feedback loop of a high gain amplifier.

Multiplying devices come in many varied forms. The most popular device for multiplying is the Quarter Square Multiplier which is based on the following relationship:

$$xy = \frac{(x+y)^2 - (x-y)^2}{4} \tag{10}$$

that is,

$$xy = \frac{x^2 + y^2 + 2xy - (x^2 - 2xy + y^2)}{4} \tag{11}$$

or

$$xy = \frac{4xy}{4} = xy. \tag{12}$$

To develop this multiplication using this technique the two variables x and y must be summed and then squared while the difference of x and y is squared and subtracted from $(x + y)^2$. One then has

$$(x + y)^2 - (x - y)^2. \tag{13}$$

If this is divided by 4 the result is

$$\frac{(x + y)^2 - (x - y)^2}{4} = xy. \tag{14}$$

Figure VI-5, indicates how this may come about using summers and a squaring device.

The squaring devices are built using diode functions similar to the ones explained in the next two sections.

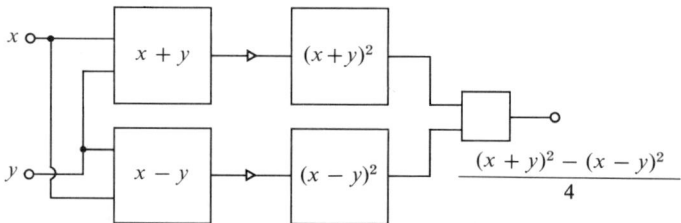

Fig. VI-5. Functional Relationships of Quarter Square Multiplier

The Generation of Arbitrary Functions (Fixed Diode Function Generator)

In using analog computers it is often necessary to generate particular types of functions. These functions may be generated using any of several common techniques including

1. power series formulation,
2. the formulation of differential equations,
3. simulation of constraints (if constraints are imposed), and
4. the use of function generators such as the Diode Function Generator (DFG).

Although all of these techniques have their place, the DFG technique is of particular importance. DFG's which are of the fixed (FDFG), or variable (VDFG) types are used frequently to generate the more familiar and more often required functions.

It is the purpose of this section to provide an understanding of how the Fixed Diode Function Generator (FDFG) performs and how it may be used effectively in computer programs.

The FDFG technique of generating functions can best be understood by considering the following circuit.

NON-LINEAR PROGRAMMING ELEMENTS AND CIRCUITS

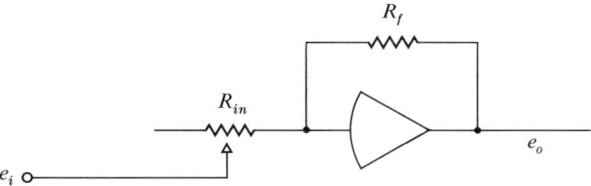

Fig. VI-6

The gain of this circuit $|e_o/e_i|$ in terms of the resistances R_f and R_{in} is

$$\left|\frac{e_o}{e_i}\right| = +\frac{R_f}{R_{in}} \qquad (15)$$

where R_{in} is variable (and $\neq 0$).

From this equation, it is evident that the ratio of $|e_o/e_i|$ may be varied during computation. One may draw the relationship between e_o and e_i where R_{in} is continuously variable as

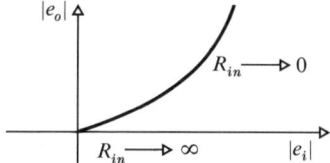

(The slope of this curve is the gain R_f/R_{in}. The max. slope is the amplifier's open loop gain.)

Fig. VI-7. Relationship of $|e_o|$ and $|e_i|$

This technique suggests that R_{in} be continuously variable to obtain the particular curve desired. Although this is possible, it is not practical. However, an approximation to the curve can be obtained by putting together a series of straight line segments. The circuit of a resistor and diode network that will produce a straight line segment is shown below.

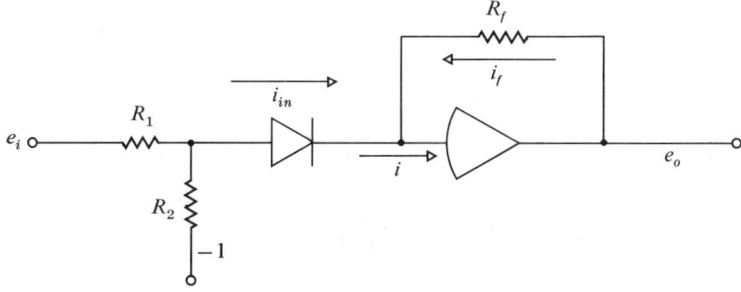

Fig. VI-8

When the diode in this circuit is not conducting, the output e_o is zero.

When the diode conducts, the currents may be summed at the input to the high-gain amplifier, yielding an output e_o which is proportional to the input e_i and the combinations of resistors R_1 and R_2 chosen. Let us look at the output for a typical resistive network at the input.

For the following configuration the output is as indicated

(a)
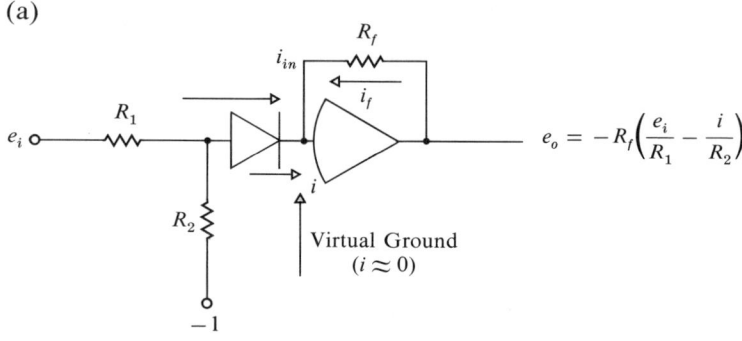

(b)
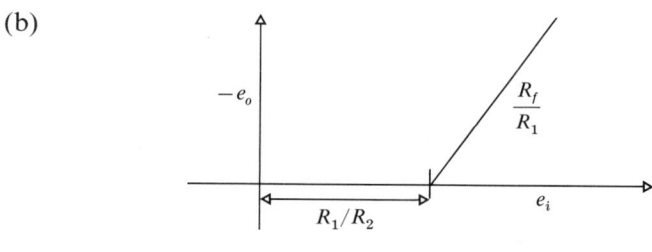

Fig. VI-9

Note that each such configuration will produce a straight line segment which begins to appear after the point $e_{in} = R_1/R_2$. The point R_1/R_2 is referred to as a *breakpoint*; R_f/R_1 is the *slope* of the line segment. Now let us see how one can use this information. For the following configuration, the straight line segments produced by closing switches SW 1 and 2 at the appropriate voltage levels are shown in Fig. VI-10b. This idea of switching in straight line segments is the basis upon which Fixed Diode Function Generators (FDFG's) operate. That is, a particular function is generated by the superposition of straight line segments.

NON-LINEAR PROGRAMMING ELEMENTS AND CIRCUITS

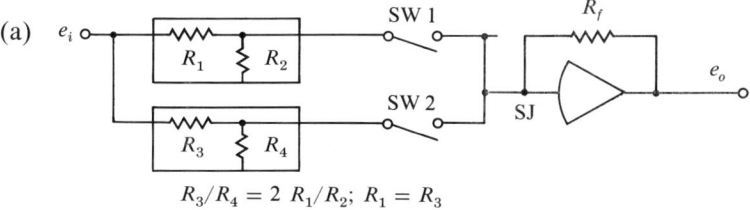

(a)

$R_3/R_4 = 2\, R_1/R_2;\ R_1 = R_3$

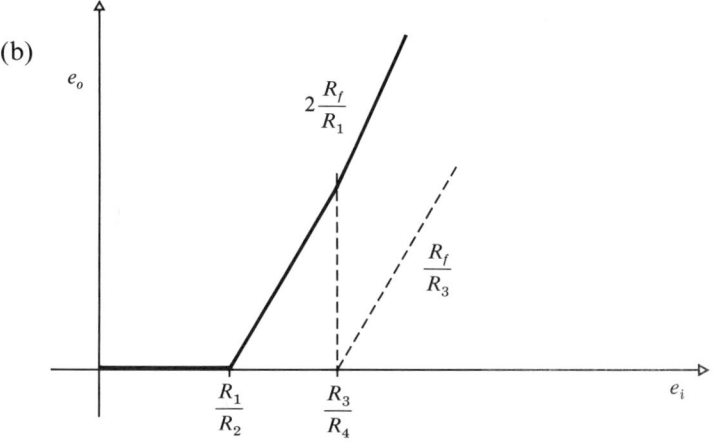

(b)

Fig. VI-10

In the diagram below, the straight line segments obtained by closing switches 1, 2, 3 and 4, at the appropriate voltage levels are shown in Fig. VI-11b. All lines have the same slope. Note the inverter in the third and fourth segments.

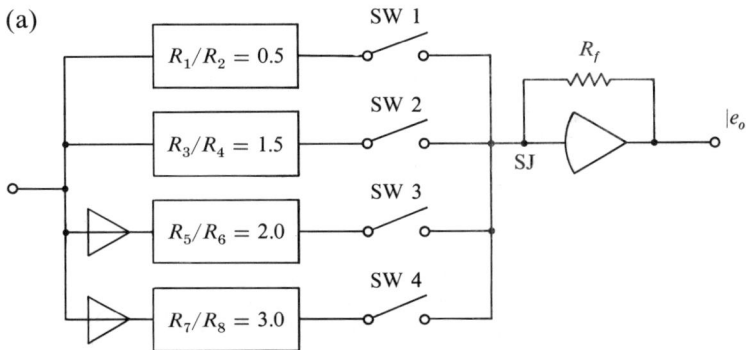

(a)

Fig. VI-11

(b)

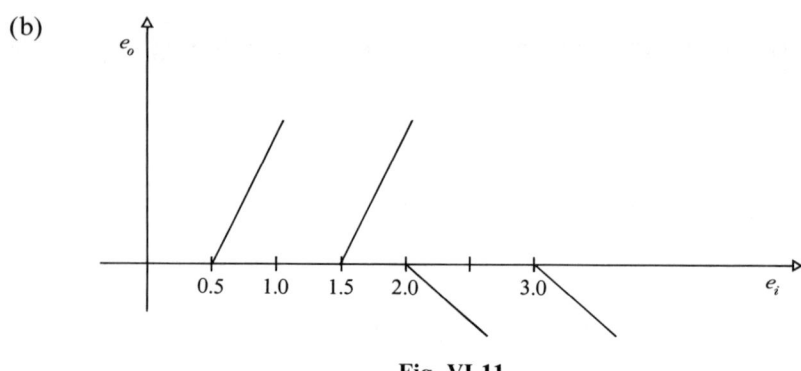

Fig. VI-11

The resulting composite configuration is indicated below.

(c)

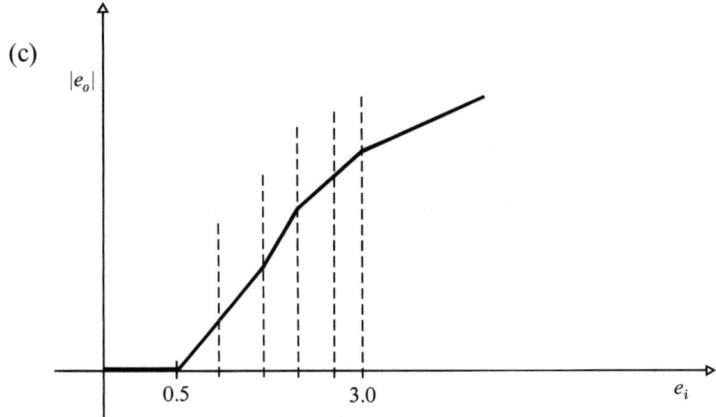

Fig. VI-11

One should now have a very good idea how FDFG's perform. Let us investigate how they may be used to generate functions for computer programs. This is really the more interesting and practical information about FDFG's.

The computer symbol for a FDFG is indicated in Fig. VI-12. Note that an external high-gain amplifier must be provided.

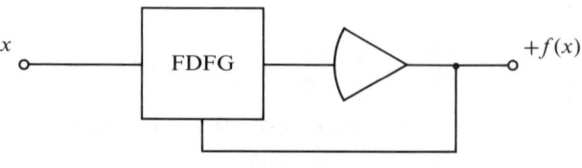

Fig. VI-12

The computer symbol for a FDFG used for generating sin x is shown in Fig. VI-13.

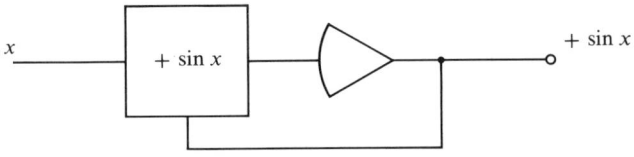

Fig. VI-13

Many companies produce FDFG's that generate certain standard functions. At the present time, FDFG's that provide the following functions are available:

$$f(x_1) = \sin x, \cos x$$
$$f(x_1) = 1/1 \log x, \log x$$
$$f(x_1) = x^2.$$

Using these FDFG's as basic building blocks, one may generate functions explicitly, or may use circuit configurations containing FDFG's to generate other functions (or hybrids of these functions). Let us see how one may use a sin x DFG in a particular program. The pendulum bob shown below is acted upon by two forces in its plane of motion—forces due to gravity and friction.

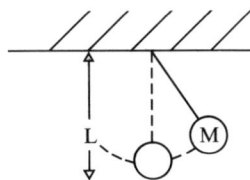

Fig. VI-14

The equation for its rotational motion in two dimensions is obtained by equating all torques about the pivot point.
This equation is

$$ML^2 \frac{d\theta^2}{dt^2} + KL\frac{d\theta}{dt} + MgL (\sin \theta) = 0. \qquad (16)$$

Solving it for the highest order derivative we obtain

$$\frac{d^2\theta}{dt^2} = -\frac{K}{ML}\frac{d\theta}{dt} - \frac{g}{L}(\sin \theta) \qquad (17)$$

where

$$M = 2 \text{ slugs} \qquad g = 32 \text{ ft./sec.}^2$$
$$L = 1/2 \text{ ft.}$$
$$K = \frac{10 \text{ lb.-ft.}}{\frac{\text{radian}}{\text{sec.}}}.$$

A trivial solution to this equation exists for very small angles of θ, angles for which $\sin \theta \simeq \theta$. However, suppose θ is larger; then, one must generate the function $f(\theta) = \sin \theta$ in obtaining the computer solution to this equation.

Two cases will be considered:

(1) $|\theta| \leq \pi$ and
(2) $|\theta| \leq \pi/4$

with any appropriate initial conditions. To evaluate the maximum values of the output of amplifiers both θ max and $\dot{\theta}$ max must be known.

Since,

$$\omega_n \approx \sqrt{\frac{g}{L}} = \sqrt{\frac{32}{1/2}} = \frac{8 \text{ rad.}}{\text{sec.}} \qquad (18)$$

$$|\theta| \leq \theta$$
$$|\dot{\theta}| \leq \dot{\theta}$$

then the scaled variables are:

P.V.	R.U.M.	C.V.
θ	π	$\left[\dfrac{\theta}{\pi}\right]$
$\dot{\theta}$	$8\pi \approx 25$	$\left[\dfrac{\dot{\theta}}{25}\right]$

The unscaled computer diagram for the solution of the pendulum equation of motion with $\dot{\theta}(0) = 0$, $\theta(0) \neq 0$ is shown below.

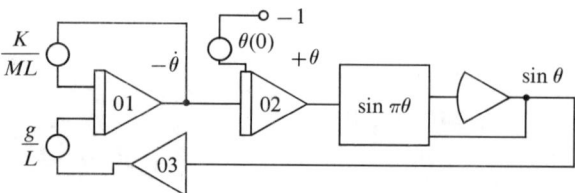

Fig. VI-15

NON-LINEAR PROGRAMMING ELEMENTS AND CIRCUITS 75

The output of amplifiers 1 and 2 in the scaled equations are:

$$A01 = -\left[\frac{\dot{\theta}}{25}\right] \tag{19}$$

$$A02 = +\left[\frac{\theta}{\pi}\right]. \tag{20}$$

The computer diagram with the scaled outputs is indicated below.

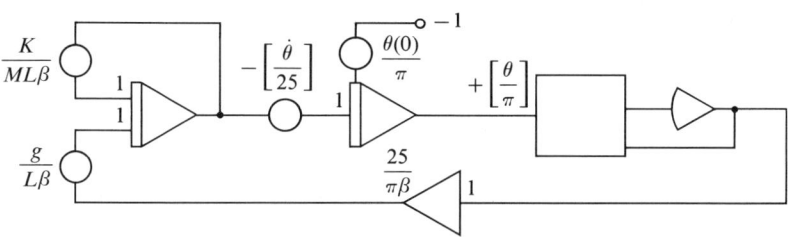

Fig. VI-16

What is the output of the function generator? Is your answer $\sin[\theta/\pi]$? Well, that is incorrect. For the type of sine generators available, you have a choice of outputs. That is, the sine/cosine FDFG will approximate the curves of any one of the following trigonometric functions with seven segments per quadrant.

$$\begin{array}{ll} \sin\theta & -180° \leq \theta \leq +180° \\ -\cos\theta & -180° \leq \theta \leq +180° \\ +\cos\theta & -180° \leq \theta \leq +180° \\ \sin\theta & -90° \leq \theta \leq +90° \end{array}$$

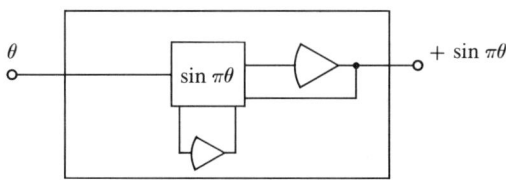

Fig. VI-17

Now, what should the output of the sine DFG be?

$$\text{OUTPUT} = \sin\theta \tag{21}$$

which is the function we want.

The completely scaled computer diagram for case I is shown in Fig. VI-18.

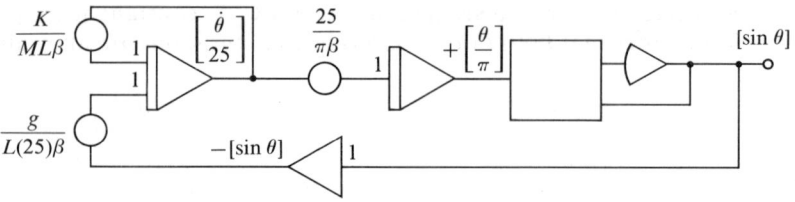

Fig. VI-18

The settings of the *pots* and the *amplifier* gains will now be indicated by choosing an appropriate β.

For $\beta = 10$

$$\beta 01 = \frac{K}{ML\beta} = \frac{10}{2(1/2)\beta} = 1 \tag{22}$$

$$\beta 02 = \frac{g}{L(25)\beta} = \frac{32}{1/2(25)10} = 0.256 \tag{23}$$

$$\beta 03 = \frac{25}{\pi\beta} = \frac{25}{\pi(10)} = 0.796. \tag{24}$$

The program has been completely sealed for θ max $= \pi$.
Now let us do it for θ max $= \pi/4$ using the same sine/cos DFG.

P.V.	R.U.M.	C.V.
θ	$\pi/4$	$\left[\dfrac{4\theta}{\pi}\right]$
$\dot{\theta}$	$\dfrac{8\pi}{4} \approx 2\pi$	$\left[\dfrac{\dot{\theta}}{2\pi}\right]$

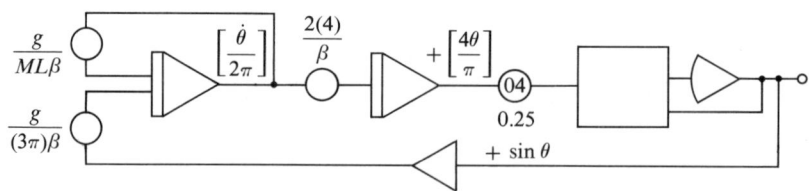

Fig. VI-19

You will notice that pot 04 had to be added to obtain the proper value at the output of the FDFG.

It should be noted, at this point, that generally one does not use potentiometers at the inputs of Diode Function Generators. However, in cases where the input resistance to the DFG is constant it may be done. Remember that both of the following configurations are allowable.

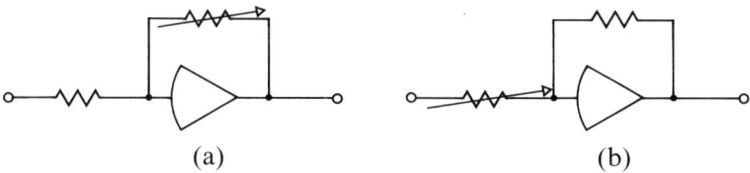

(a)   (b)

Fig. VI-20

The SINX DFG has the nonlinear resistor in the feedback path and may be fed from a pot.

From this example, one can see that in order to use FDFG's they must be scaled properly. Therefore, whenever using a FDFG make certain that you are indeed generating the function desired.

Let us look at another example involving a function log 10x. Suppose we have the differential equation

$$a\ddot{x} + b\dot{x} + cx = f(x)$$

where $f(x)$ represents a forcing function on the second order system described by this equation. Solving the equation for the highest order derivative one gets:

$$\ddot{x} = -\frac{b}{a}\dot{x} - \frac{c}{a}x + \frac{1}{a}f(x). \qquad (25)$$

Draw an unscaled computer diagram that will enable you to solve this equation.

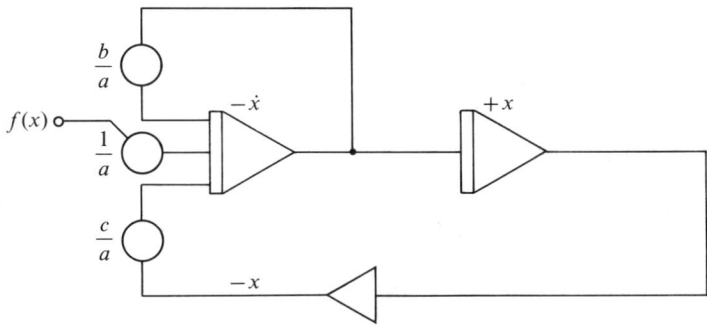

Fig. VI-21

It will be noticed that $f(x)$ appears explicitly on the input of the first integrator.

Our next step is to determine explicitly the function $f(x)$, and using the computer reference handbook, determine how we can generate this function. Assume that $f(x) = \log 10x$ and that our reference manual indicates that we can generate $-1/2 \log(100x)$, in unit scaling, by using the following configuration:

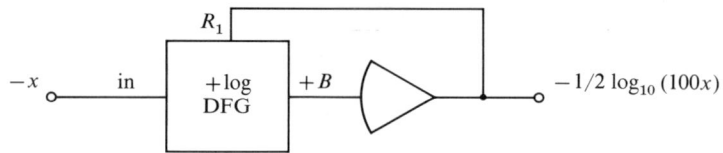

Fig. VI-22

To solve this problem $-1/2 \log_{10}(100x) = f'_1(x)$ must be expanded. You may try this yourself or look at the way we have done it.

$$f'_1(x) = -1/2(\log_{10}100 + \log_{10}x) \qquad (26)$$
$$f'_1(x) = -1/2(2 + \log_{10}x) \qquad (27)$$
$$f'_1(x) = -1-1/2 \log_{10}x \qquad (28)$$

Can you think of a way of generating $f'_1(x)$? Try one!

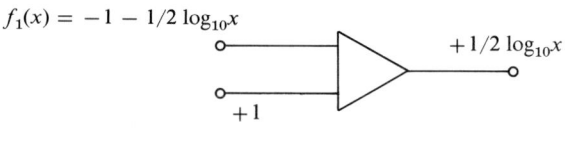

Fig. VI-23

Combine the log DFG with the additional circuitry to produce

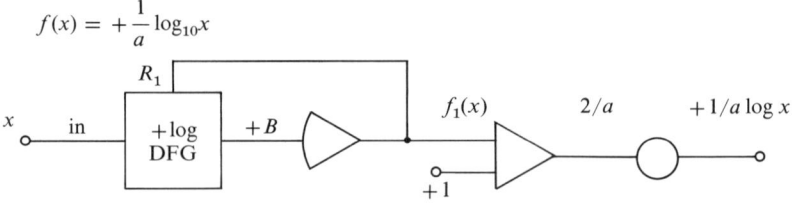

Fig. VI-24

Notice that in addition to the log x, DFG circuitry was required to produce the needed function $f(x)$. This is usually the case, although

FDFG's exist they must be implemented to the particular problem at hand.

SUMMARY

This material presents the basic technique of generating functions using Diode Function Generators and specifically stresses the use of Fixed Diode Function Generators (FDFG's) such as $\sin x$ and $\log x$. The important point stressed was that the FDFG's must be adapted to the particular problem being solved. That is, scaling for each problem must be considered.

Other methods are employed for generating poorly defined functions. In cases where the particular function to be generated is not explicitly defined or does not have a FDFG developed for it, one must use Variable Diode Function Generators (VDFG). In these devices one has control of the breaking points and slopes of the individual straight line segments.

The Mechanization of Constraints and Elementary Logic Functions

Some of the more effective ways to generate nonlinear characteristics, or impose system constraints in a given computer problem involve the use of amplifiers with diodes and relay switches.

Diodes are semiconductor or thermionic devices that conduct essentially in only one direction—in the forward direction (as indicated by the arrow in the symbol). For the present we shall assume that the diode is an "ideal" switch, characterized by a current of zero when reverse voltage is applied and by a zero voltage drop across it, when forward current flows.

The symbols for a diode and the analogous "ideal" switch are shown below.

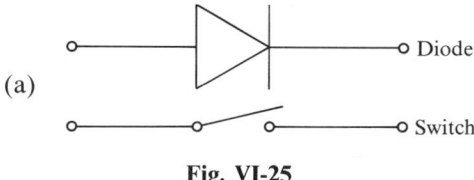

Fig. VI-25

Draw the switch showing its position (open or closed) for the case where a voltage is applied to the diode as shown in the diagram below.

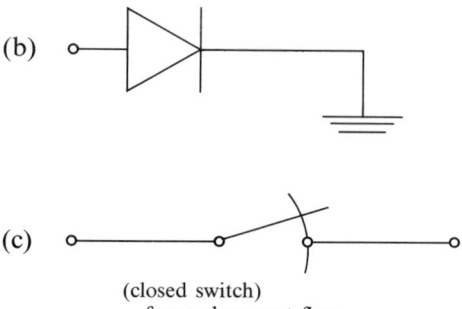

Fig. VI-25

Reverse the polarity of the voltage applied to the above diode and indicate the analogous switch:

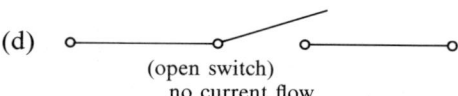

Fig. VI-25

Since diodes are essentially switches that are closed when forward biased and open when reversed biased, they can be used to limit the voltage at the output of a computer amplifier. When used as limiters they are connected either in series with the current paths to the amplifier inputs or in parallel with them.

Consider the inverter amplifier indicated below:

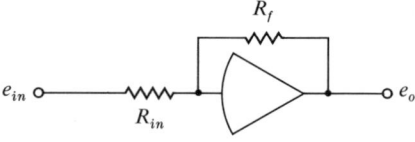

Fig. VI-26

The output voltage of this amplifier is

$$e_o = -\left(\frac{R_f}{R_{in}}\right)e_{in}. \tag{29}$$

The characteristic of the inverter amplifier for both positive and negative input voltages is indicated below.

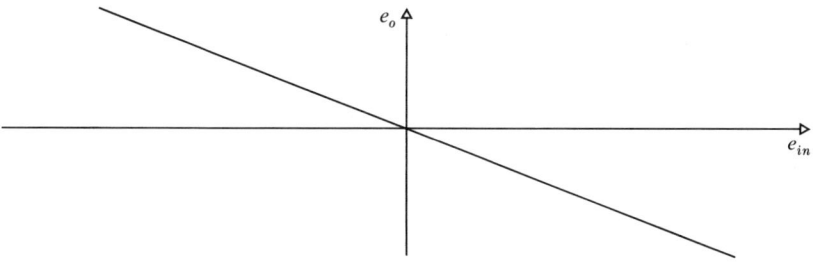

Fig. VI-27

What is the slope of this characteristic?

$$\text{Slope} = -\frac{R_f}{R_{in}}. \tag{30}$$

What is the output of the circuit below?

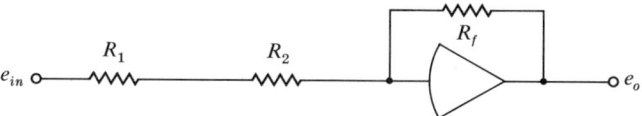

Fig. VI-28

$$e_o = -\left(\frac{R_f}{R_1 + R_2}\right)e_{in}. \tag{31}$$

Assume that the configuration of the circuit is changed to appear as below:

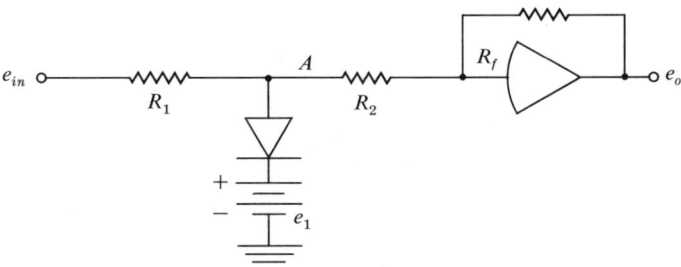

Fig. VI-29

When point A reaches $+e_1$ volts, what is the input to the amplifier?

$$\text{input} = +e_1. \tag{32}$$

The voltage at point A can't rise above $+e_1$ because the voltage across the diode is zero in this forward (conducting) state. Now, what is the voltage at the amplifier output?

$$\text{output} = e_o = -\left(\frac{R_f}{R_2}\right)e_1. \tag{33}$$

It appears that the output voltage remains at a constant level after e_{in} reaches a certain input level. Let us see if we can determine this input voltage level.

The voltage across resistors R_1 and R_2 is e_{in}. Since R_1 and R_2 represents essentially a voltage divider, the voltage across resistor R_2 (the voltage which determines when the diode switches) is

$$\left(\frac{R_2}{R_1 + R_2}\right)e_{in} = e_1. \tag{34}$$

Therefore, solving for e_{in}

$$e_{in} = e_1\left(\frac{R_1 + R_2}{R_2}\right) \tag{35}$$

one can determine the level of input voltage for which the output voltage becomes constant.

The characteristics of this circuit are shown in Fig. VI-30.

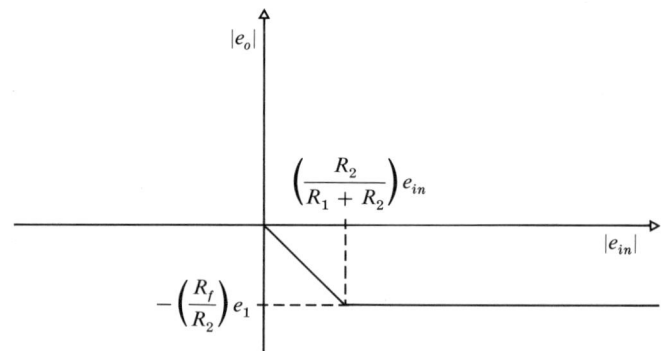

Fig. VI-30

The circuit configuration to obtain the following characteristic is shown in Fig. VI-31.

NON-LINEAR PROGRAMMING ELEMENTS AND CIRCUITS

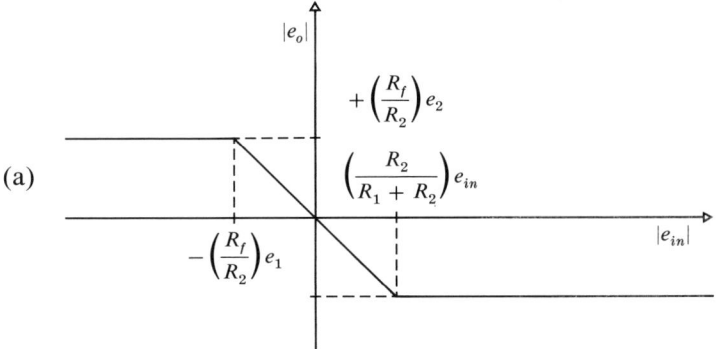

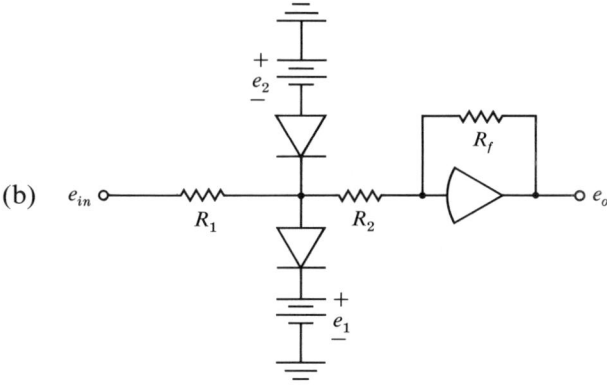

Fig. VI-31

What is the output (e_o) at the second switching voltage e_2?

$$e_o = +\left(\frac{R_f}{R_2}\right)e_2. \tag{36}$$

What are the values of input voltage e_{in} at these switching points?
The diode connected to voltage e_1 will switch when the voltage at its positive terminal is e_1. That is,

$$e_1 = \left(\frac{R_2}{R_1 + R_2}\right)e_{in} \tag{37}$$

or,

$$e_{in} = \left(\frac{R_1 + R_2}{R_2}\right)e_1. \tag{38}$$

At the other switching point,

$$e_2 = -\left(\frac{R_2}{R_1 + R_2}\right)e_{in} \tag{39}$$

or,

$$e_{in} = -\left(\frac{R_1 + R_2}{R_2}\right)e_1. \tag{40}$$

This suggests that one can adjust the value of input voltage at which limiting will take place by varying the resistances in the input circuit. Up to now you have learned what diodes are and how limiting can be performed with amplifiers. Let us see what other interesting things can be done.

Another circuit that will provide the same characteristic is shown in Fig. VI-32. It consists of two diodes, each in series with a reference voltage and connected in parallel with the amplifier.

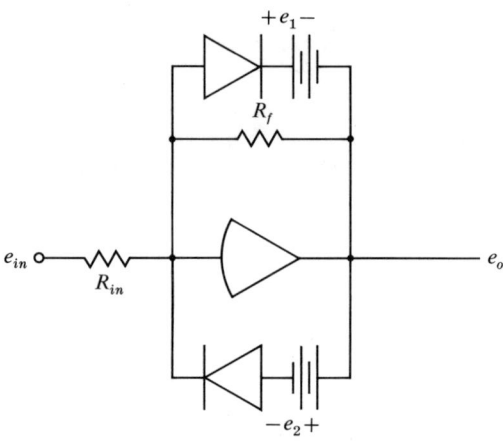

Fig. VI-32

This circuit is referred to as the feedback or "soft" limiter. The input voltage levels beyond which the output voltage remains constant are

$$e_{in} = \left(\frac{R_{in}}{R_f}\right)e_1 \tag{41}$$

and,

$$e_{in} = -\left(\frac{R_{in}}{R_f}\right)e_2. \tag{42}$$

NON-LINEAR PROGRAMMING ELEMENTS AND CIRCUITS 85

The characteristic of this circuit indicating the value of the slope and switching points is shown in Fig. VI-33.

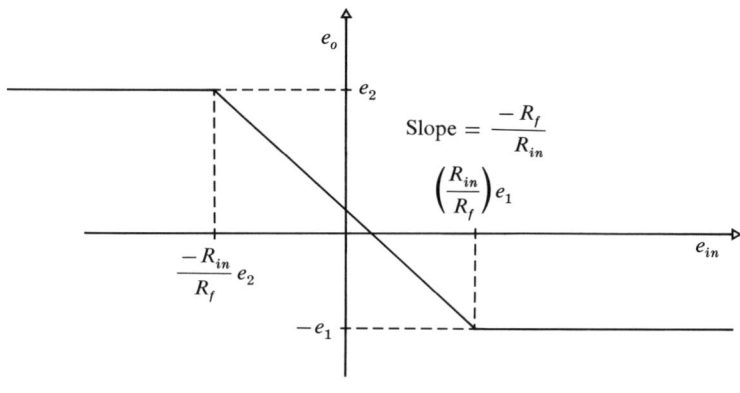

Fig. VI-33

If $R_{in} = R_f$ and the voltage sources are represented by potentiometers connected to the computer reference voltage, the circuit configuration is indicated in Fig. VI-34.

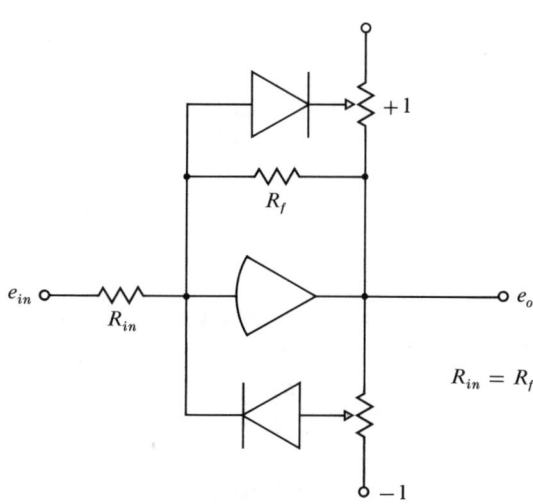

Fig. VI-34

This configuration using the summing amplifier symbol instead is shown below.

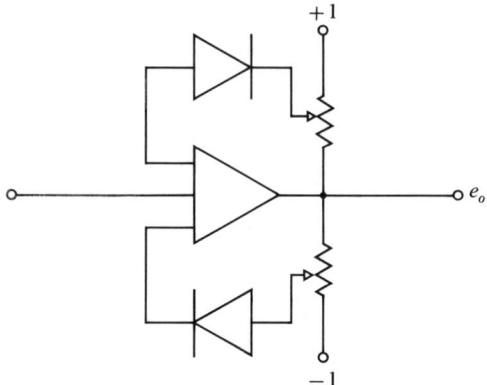

Fig. VI-35

Between the limits, e_1 and e_2, neither diode will conduct if potentiometers 1 and 2 have been properly set. When, for example, the output voltage reaches a value slightly above the upper limit ($e_o > e_2$), diode 2 conducts. This places a low resistance in parallel with the normal feedback resistor of the amplifier. Thus, the gain of the amplifier is reduced by a factor of 100 or more. The other diode cannot conduct when e_o is positive because it has a reverse (negative) voltage across it.

The potentiometer settings for a specified limit can be estimated by considering the input-output relations of potentiometers. The potentiometer settings for either limit are:

$$K_1 = \frac{|e_1|}{1 + |e_1|} \quad (43)$$

$$K_2 = \frac{|e_2|}{1 + |e_2|} \quad (44)$$

Let us see how these limiter circuits can be used to provide actual constraints in a problem on the computer. This is an interesting and valuable example.

Consider the problem of obtaining a characteristic which goes between two limits, namely:

$$\begin{aligned} e_o &= e_1 \text{ if } e_{in} < e_p \\ e_o &= -e_2 \text{ if } e_{in} > e_p. \end{aligned} \quad (45)$$

The characteristic for this limiting is shown below.

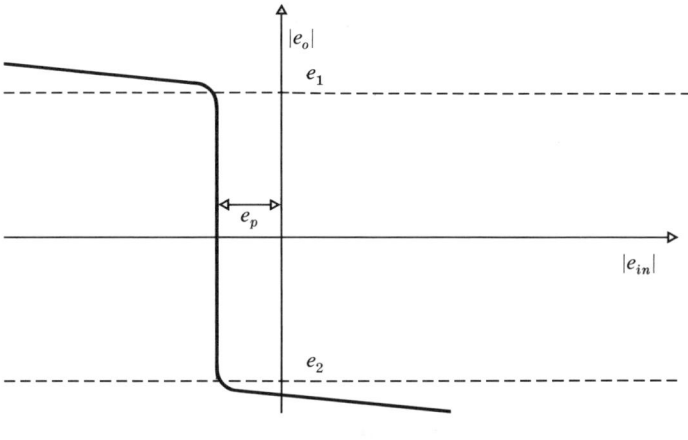

Fig. VI-36

Now, we must determine how such a characteristic may be generated! The value of the slope of the function when limiting is not occurring is (Slope = 0).

This is interesting! Obviously then, part of the desired characteristic may be obtained by using a feedback limiter with the feedback resistor R_f equal to zero. This circuit configuration is shown in Fig. VI-37.

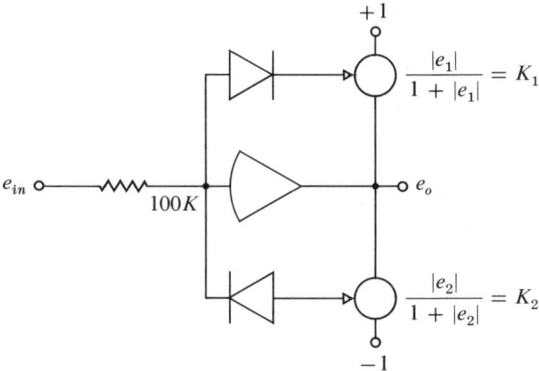

Fig. VI-37

For $e_1 = -0.2$ mu and $e_2 = -0.4$ mu* the potentiometer settings on this configuration are:

$$K_1 = \frac{|0.2|}{1 + |0.2|} = 0.167 \qquad (46)$$

$$K_2 = \frac{|0.4|}{1 + |0.4|} = 0.286. \qquad (47)$$

The characteristic obtained for this circuit is

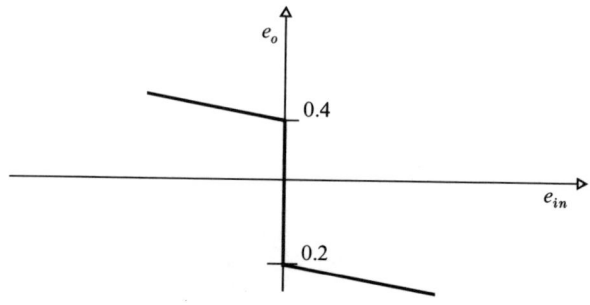

Fig. VI-38

This is not quite the characteristic we want. This whole curve must be shifted to the left by an amount e_p.

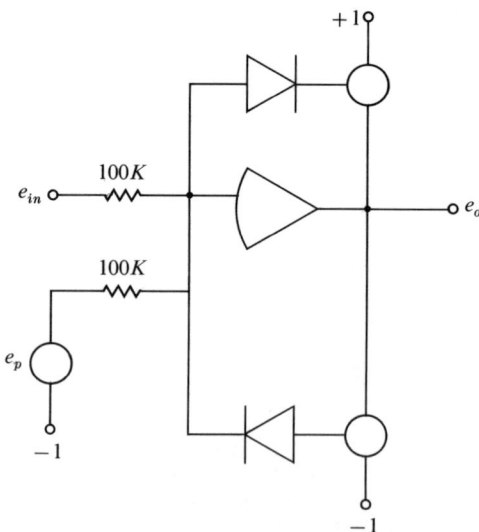

Fig. VI-39

---

*mu refers to machine units.

This shift may be obtained by adding an input $-e_p$ to the high gain amplifier. The resulting amplifier configuration that will give one this result is shown in Fig. VI-39.

This circuit is called a "BANG-BANG" circuit. The resulting characteristic of this circuit is shown in Fig. VI-40.

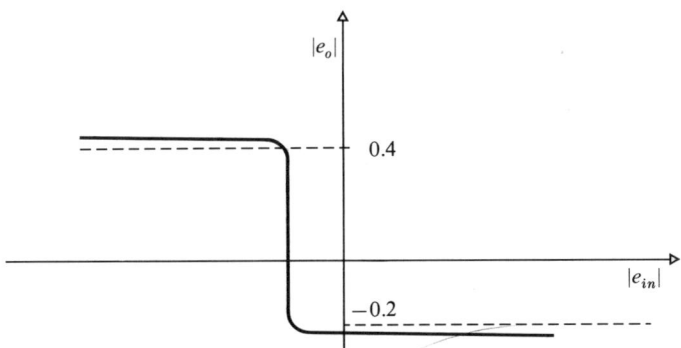

Fig. VI-40. Characteristics of Bang-Bang Circuit

Suppose it was desired to obtain the characteristic below.

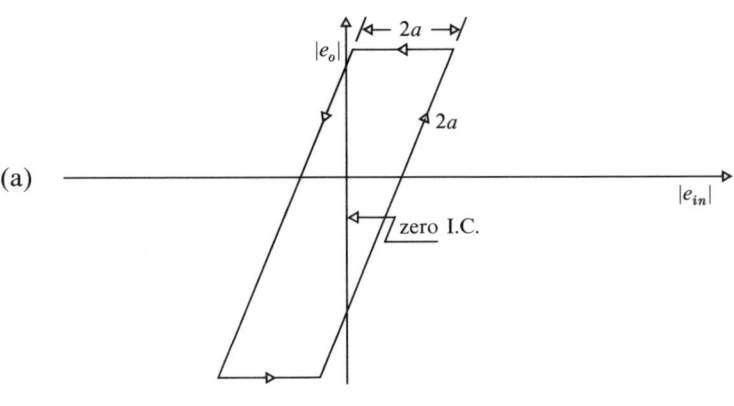

Fig. VI-41

How would one go about this? There is no need to develop a configuration to produce this characteristic. It has already been done. Refer to Appendix C of the *EAI Handbook of Analog Computation*. A number of characteristics have been simulated. The configuration that will produce the above characteristic is shown in Fig. VI-41b.

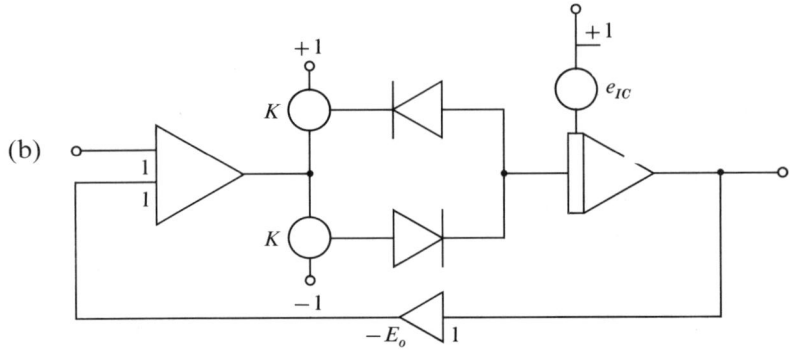

Fig. VI-41

One can now conclude that if a particular system constraint is required the procedure to be taken to obtain the desired result is as follows.

Look in Appendix C of the *EAI Handbook of Analog Computation* to determine whether or not a circuit configuration has been developed. If an appropriate circuit is not listed you are left to your own ingenuity to develop one.

Perhaps the function to be simulated is the following:

$$e_o = K_1 \text{ for } e_{in} < 0 \qquad (48)$$
$$e_o = K_2 \text{ for } e_{in} > 0 \qquad (49)$$

How can this be accomplished? If your answer is to use the feedback limiter, you are correct, but there is another method.

Suppose ordinary switches are used in conjunction with potentiometers. You could accomplish the simulation using the configuration in Fig. VI-42.

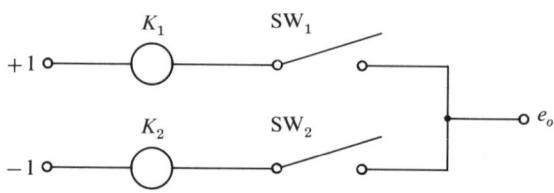

Fig. VI-42

When $e_{in} < 0$ close switch SW 1 and when $e_{in} > 0$ open switch SW 1 and close switch SW 2. This will give the following characteristic:

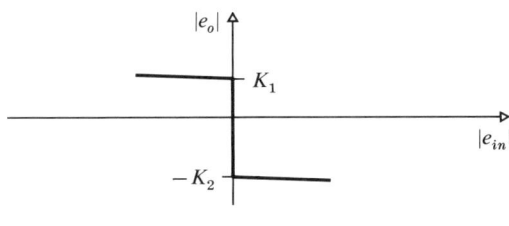

Fig. VI-43

This suggests that perhaps nonlinear constraints can be obtained by using relay switches. Fortunately, the analog computer has devices which automatically *open* and *close switches*. They are called comparators.

A simplified programmer's symbol for the comparator is

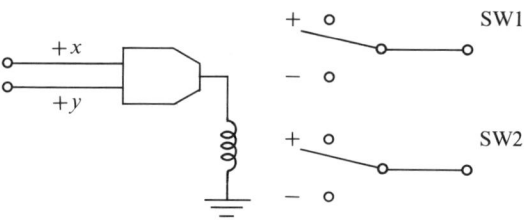

Fig. VI-44

Indicate by an equation when switches 1 and 2 will be on the positive terminals:
$$x + y > 0. \tag{50}$$
If $y = 0$ volts, what polarity must x have for switching to take place?
$$x \geq 0 \text{ volts.} \tag{51}$$
Now, consider using comparators to solve the problem that was solved using *diodes* above, namely
$$e_o = K_1, \ e_{in} < 0 \tag{52}$$
$$e_o = K_2, \ e_{in} < 0 \tag{53}$$

The comparator configuration required to do this is shown in Fig. VI-45.

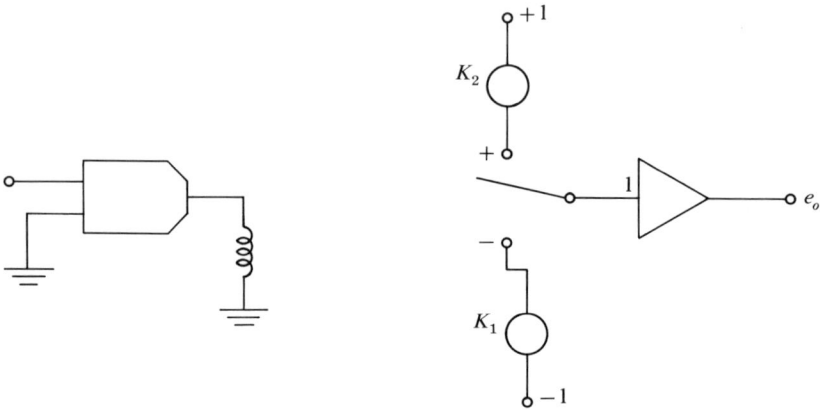

Fig. VI-45

What comparator configuration is required to obtain the following constraint? This involves an application of the concepts considered previously.

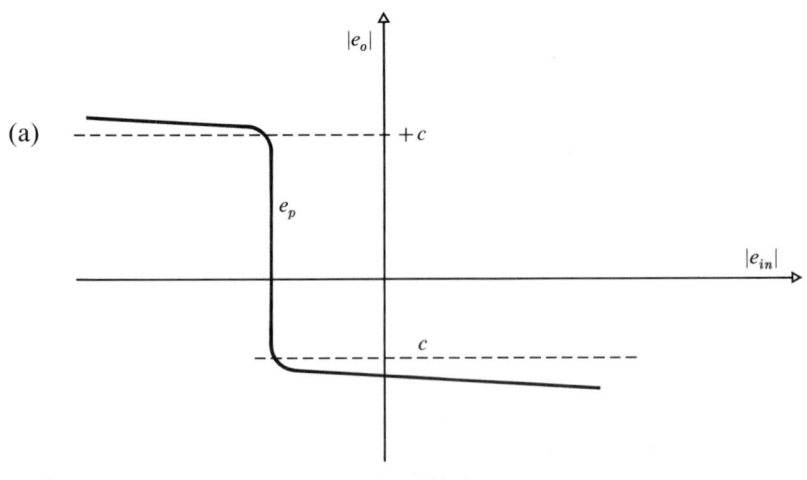

Fig. VI-46

Answer:

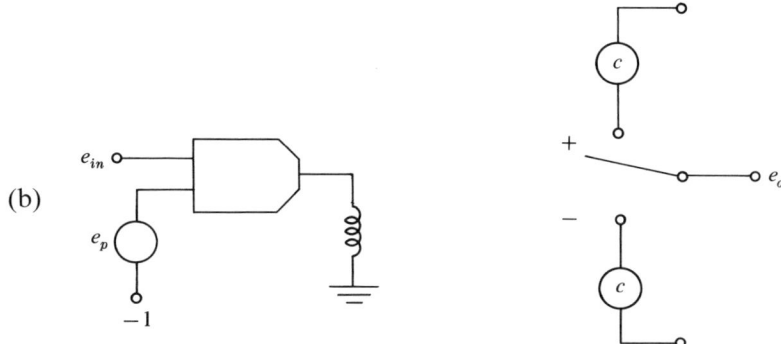

(b)

Fig. VI-46

This is interesting! It seems that both of the constraints developed previously can be simulated just as well using relay comparator circuits.

Look in Appendix C of the *EAI Handbook of Analog Computation* to determine the relay configuration required to simulate the following characteristic:

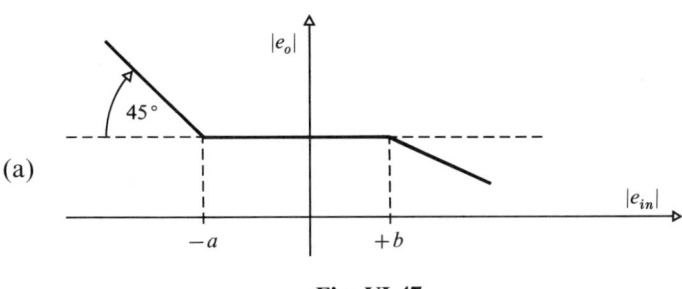

(a)

Fig. VI-47

The circuit configuration used to simulate this characteristic is indicated in Fig. VI-47b.

(b)

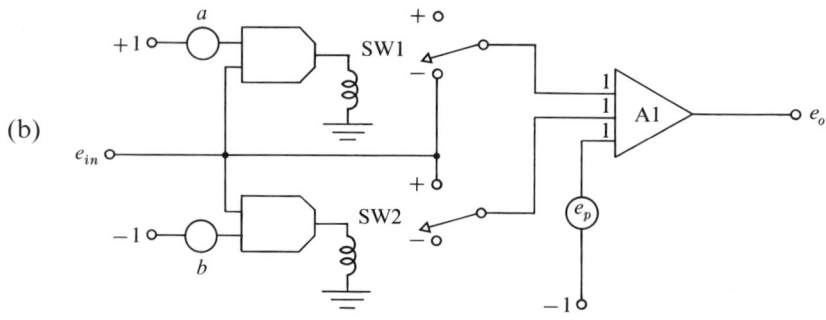

Fig. VI-47

When $e_{in} = 0$, what is the output of amplifier 1?

$$e_o = +e_p. \tag{54}$$

At what input voltage e_{in} will switch (SW 2) go to the positive terminal?

$$e_{in} \geq b. \tag{55}$$

What is the output of amplifier 1 in this position?

$$e_o = -(-e_p + e_{in})$$
$$e_o = -(-e_p + b)$$
$$e_o = +e_p - b. \tag{56}$$

In which position is SW 1 at this input voltage (b)?

SW 1 is on positive terminal 1, since sum of inputs is still positive. At what value of input e_{in} will SW 1 switch to negative terminal?

$$e_{in} = -a. \tag{57}$$

What is value of output at this input potential?

$$e_o = -(-e_p - a)$$
$$e_o = e_p + a. \tag{58}$$

One will note that this configuration indeed satisfies the constraint characteristic indicated above.

Congratulations! You have finished another chapter. This chapter indicated two ways of obtaining nonlinear characteristics of system constraints. The first method utilized unilateral devices, namely diodes, in conjunction with amplifiers and resistors. The second method utilized relay switches, namely comparators. Both techniques may be used to generate characteristics of desired system constraint. When simulating a nonlinear function, choose that circuit which will give you the best results at the least cost of circuit components.

Conclusion

This chapter reviewed the basic concepts of non-linear programming using analog computers. In particular, it concerned itself with non-linear component configurations which perform the following functions:

(1) multiplication and division of variables,
(2) the generation of arbitrary functions, and
(3) the mechanization of constraints and elementary logic operations.

This chapter has been written in such a way that one should be able to use non-linear components in programming configurations without much difficulty.

References

1. EAI, *Handbook of Analog Computation* (Princeton, New Jersey).
2. Korn & Korn, *Electronic Analog and Hybrid Computers* (New York: McGraw-Hill Book Company, 1964).
3. S. R. Ledley, *Digital Computer and Control Engineering* (New York: McGraw-Hill Book Company, 1960).

CHAPTER VII

Analog/Logic Techniques and Interfacing (Hybrid Computer)

MANY PROBLEMS can be solved on the analog computer as a stand alone computer. However, many times the particular problem is of such a nature that a need arises to perform logical functions in the process of solution. It is the intent of this chapter to discuss how and why an analog computer can be connected to parallel logic and what components are used. The hybrid computer to be discussed is not a digital computer connected to an analog computer but rather an analog computer connected to logic components.

The need for solving a problem on a hybrid computer arises when any combination of the following functions have to be performed.

(1) storage
(2) counting
(3) updating
(4) logical decisions
(5) tracking
(6) comparing
(7) digital to analog and analog to digital conversion
(8) automatic control

To appreciate the advantages which hybrid computation offers, it is necessary to understand the hybrid system, including the analog computer and patchable logic. The major purpose of the analog computer is for high speed solution of ordinary differential equations. Since the dynamics of most physical systems are defined by sets of differential equations, this function is essential in nearly all applications. When the physical system is described by a set of partial differential equations, the PDE's are reduced to ODE's before solution on the analog computer. In addition to this basic mathematical function, the modern analog computer can simulate any type of discontinuities and logical function which may be present in the physical system.

To increase the efficiency of an analog computer and eliminate man interceding in the running of the computer, the analyst is replaced with a combined analog and logic program which permits automatic execution of a planned sequence of runs. The general block diagram for this type of program is shown in Fig. VII-1.

The analog computer solves the system equations at high speed, while the logic executes the runs, modifies input conditions and controls the display devices. If an interation criterion can be established, the input conditions can be automatically adjusted to satisfy this criterion.

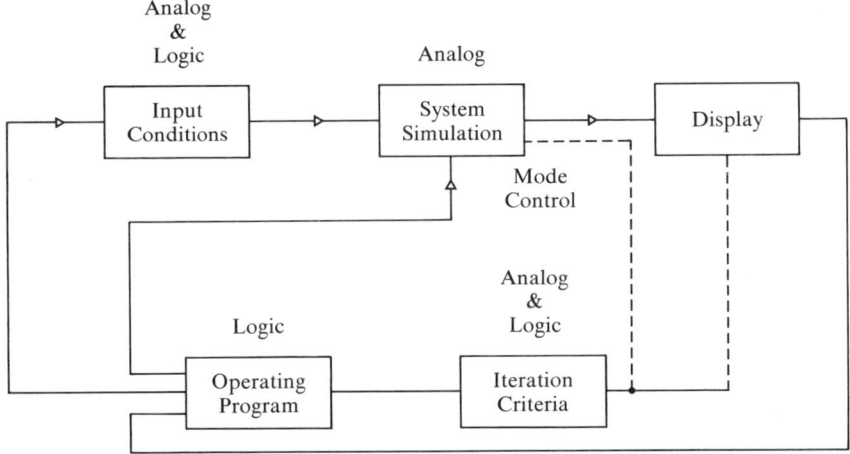

Fig. VII-1. Analog Computer Automation

Digital logic in a *hybrid* system is also used for the simulation of that portion of a system which is described as a sequence of logical operations rather than mathematical equations. In this case, the logic elements perform logical operations much the same as the analog elements perform continuous operations.

Patchable logic is also used to perform certain types of data reduction of analog results such as probability distributions, frequency measurements, peak sampling, etc. This can result in a significant reduction in the load on the digital computer as well as permitting high frequencies of information to be handled since the parallel logic is faster than the digital computer for these operations.

Finally, the patchable logic can be programmed to serve as a flexible time base for a combined hybrid problem. Since the logic section can communicate conveniently with both the analog and digital domain, it is most effective at controlling the overall problem execution.

Before analyzing the functions of the hybrid computer, it may be of interest to establish the general structure of a modern hybrid computing system. Figure VII-2 shows the configuration of the analog/logic computer which consists of four sections:

1. parallel analog computing components
2. parallel logic components
3. analog logic interface components
4. system monitoring and control.

The analog component inputs and outputs, the analog terminations for the logical interface and inputs to the analog display devices are brought to one section of the patch panel. The parallel logic components, the logical terminations for the interface, and the control of the display devices are terminated on a second section of the same patch panel. Monitoring and control includes the pot setting system, component selection and readout through the DVM, analog and logic mode control, digital logic readout and control, and the master mode control timer.

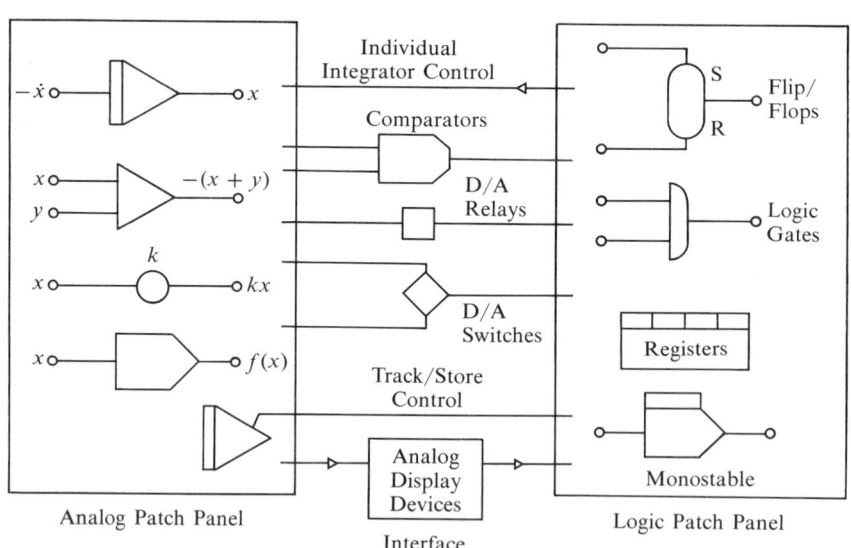

Fig. VII-2. Analog/Logic Computing System

The analog section of the hybrid computer provides a balanced set of high accuracy wide bandwidth analog computing elements. The logic components provide for both combinatorial logic by means of universal logic gates and sequential synchronous logic by means of flip-flops, shift registers, and counters. In addition, monostables are provided to fulfill a variety of timing requirements. High speed comparators are included to sense threshold conditions in the analog problem and produce logic signals. A variety of means are provided for controlling the analog simulation from logic signals, including individual integration mode and time scale controls, D/A relays which produce contact closures in the analog section, D/A switches which can switch analog signals at high speed, and individual control of track/store amplifiers.

The analog portions of the hybrid computer have been considered in previous chapters, the remainder of this chapter will be devoted to discussing interfacing components and techniques. The remaining two chapters will be concerned with the logic patchpanel and programming techniques respectively.

First of all the mode control and track store units will be considered. Then the comparator circuit will be discussed after which the D/A switch and D/A relay are considered.

Mode Control of Analog Computer

The mode of modern analog computers can be controlled by logic signals. All integrators are controllable from the master console or they also respond to signals patched directly to each one of them. Thus the whole console may be in Operate mode and one integrator (or one group of integrators) may be in IC and another group may be in Hold mode. Since a logic level has only two states it is not possible to control all three modes (IC, Hold and Operate) with a single logic signal. Two logic input terminals must be provided for this purpose.

Figure VII-3a summarizes the control of an integrator by two signals (IC) and (OP). The modes of an integrator could be written as

$$\text{Initial Condition mode} = (IC)$$
$$\text{Hold mode} = (\overline{IC}) \cdot (\overline{OP})$$
$$\text{Operate mode} = (\overline{IC}) \cdot (OP).$$

The first equation indicates that if the (IC) input is high the integrator is in the Initial Condition mode regardless of the state of the OP input. The other two modes require that the (IC) input is low.

The track/store unit which is the basic analog memory device is also under logic control. When the logic input (T) is a ONE the amplifier tracks the input r (see Fig. VII-3b) resulting in an output of $-r$. When the logic input switches to ZERO the unit switches to the STORE (or HOLD) mode and holds its output constant. Thus it remembers the last value prior to switching.

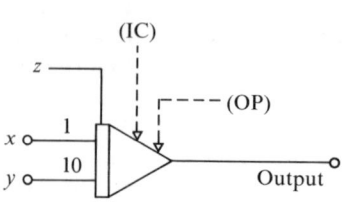
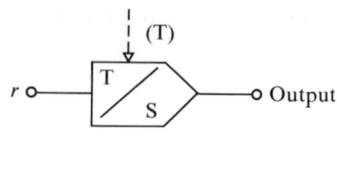

(IC) Input	(OP) Input	MODE	Output
0	0	HOLD	Hold last value
0	1	OP	$-(x + 10y)\, dt$
1	0	IC	$-z$
1	1	IC	$-z$

(T) Input	MODE	Output
1	TRACK	$-r$
0	STORE	Hold last value

(a) Integrator

(b) Track/Store

Fig. VII-3. Amplifier Control

Example 1: Consider the circuit shown on Fig. VII-4. In this circuit two different track/store type memory devices were chosen for back-to-back operation. When the track/store amplifier is in TRACK mode the integrator is in HOLD and vice versa.

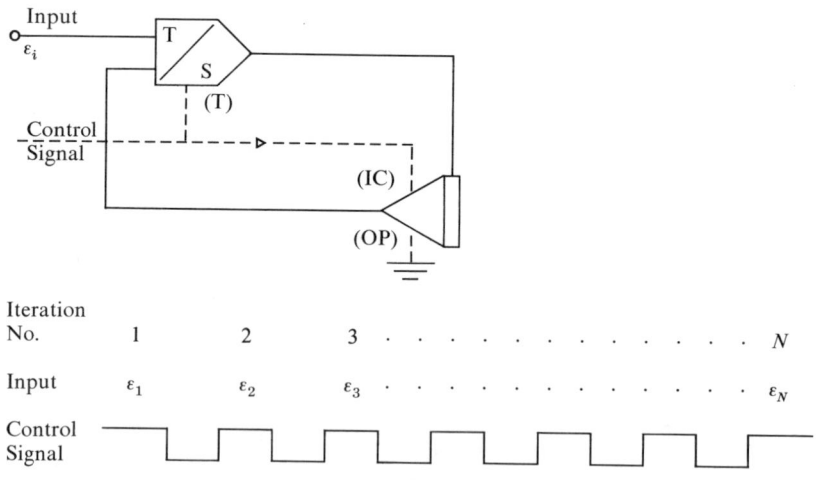

Fig. VII-4. Accumulator

Refer to the timing chart of Fig. VII-4 for the following discussion. Assume that both amplifiers start at zero. During the first iteration the T/S is tracking, resulting in -1 at its output and the integrator is in

HOLD. When the CONTROL SIGNAL reverses its state, the integrator is forced into IC mode and has an output of $+1$ while the T/S is holding -1. Then the T/S goes back to TRACK in the 2nd iteration it is tracking 1 from the integrator *and* 2 from the input. The process repeats itself until the *n*th iteration at which the T/S output becomes

$$\varepsilon_1 + \varepsilon_2 + \varepsilon_3 \cdots + \varepsilon_N = \sum_{i=1}^{N} i$$

thus the circuit "accumulates" the input ε_i.

This circuit could be used to count the number of operate cycles in a rep-op problem. Let the logic control signal be the same as the operate logic signal for the rest of the problem and the analog input be $+0.01$ reference. Then the output voltage of the integrator would be the number of completed operate cycles times 0.01 reference. If a logic ONE signal is desired after 100 operate cycles it can be obtained using a comparator with analog inputs from the integrator and from the negative reference.

The Comparator

The comparator is designed to accept analog inputs and produces a logic ONE output when the sum of the inputs is positive. When the sum of the inputs is negative the output is a logic ZERO. When the sum of the input is in the neighborhood of zero, say a small value ε away from zero, then the comparator output may be at "0" or may be at "1" dependent upon past history. This hysteresis region characterizes many "threshold" devices where a signal is compared against another. Usually ε is so small, .02% of reference, to be noticed by the operator. The output of the comparator can be latched and held constant by putting a logic ONE on the latch input. The comparator is the basic device for analog-to-logic communication.

The comparator is a synchronous device. The meaning and implication of being synchronous is discussed in a later section.

The symbol and truth table for a comparator is shown in Fig. VII-5.

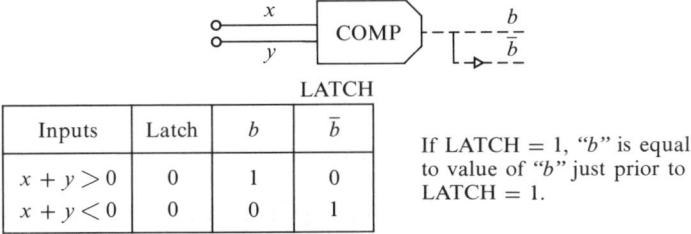

Fig. VII-5. The Comparator

The D/A Switch

The D/A switch (Digitally-controlled Analog switch) is a device that uses a logic input to control the on-off switching of an analog input to an amplifier. A logic ONE turns the switch on; a logic ZERO turns the switch off. Functionally the device acts like a resistor in series with a high speed switch that switches between the summing junction of an amplifier ("on") and ground ("off"). The size of the resistor is usually such as to provide a gain of 10 when the normal feedback resistor is used with the amplifier. The symbol for a D/A switch is a diamond as shown in Fig. VII-6. The reader is urged to prove that:

$$z = |x - y|. \qquad (1)$$

Note that if x and $-y$ happen to have the same sign, A10 may overload and the diagram should be modified slightly to avoid saturation.

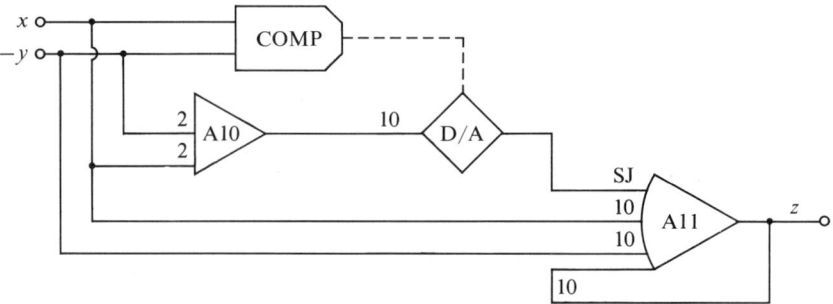

Fig. VII-6. A Circuit Using D/A Switch

The D/A Relay

The D/A relay is double throw relay, the arm of which is positioned by an input logic signal. A logic ONE positions the relay arm to analog terminal "1" and a logic ZERO to analog terminal "0."

The symbol for the D/A switch is shown in Fig. VII-7. Unlike the old-fashioned relay comparator the D/A relay does not have to be controlled by a comparator. It obeys any logic signal patched to its control and it does not care where the signal comes from. One application for the D/A switch is to change resistor sizes for automatic rescaling of an analog program in the middle of a run. Even though the switching may involve many relays the whole operation may be controlled by a single logic signal.

Fig. VII-7. D/A Relay

Conclusion

This chapter was concerned with a description of the Analog/Logic computer and its functions. It should be pointed out that the ability of analog computers to be controlled by logic elements is a direct function of the interfacing components.

The interfacing components discussed in this chapter were (1) mode control, (2) track store, (3) comparator, (4) D/A switch and D/A relay. Many examples were given to show the operation and function of these components. In Chapters VIII and IX further use of these components will be made.

References

1. EAI, DES-30, *Digital Expansion System Reference Manual,* March 1966.
2. EAI, *Basics of Parallel Hybrid Computers,* written by G. Hannauer, 1967.
3. EAI, *Handbook of Analog Computation,* Princeton, New Jersey.
4. Korn & Korn, *Electronic Analog and Hybrid Computer* (New York: McGraw-Hill Book Company, 1964).

CHAPTER VIII

Parallel Logic Components and Hybrid Programming

THIS CHAPTER intends to show how one can extend the capabilities of the analog computer by the addition of parallel logic.

The use of the analog computer in teaching engineering and science has become increasingly widespread. It is accepted as a means for providing classroom demonstrations and as a tool for independent research. The reason is that the analog computer can easily solve differential equations, linear and non-linear, which characterize so many physical systems. However, when the mathematical model requires decision making, logic, switching, timing, memory, or individual mode control of analog components, the need to add parallel logic to the analog computer becomes self-evident.

Switching devices, sampled data systems, and adaptive control systems are simulated in a natural manner using parallel logic in conjunction with analog components. In simulating a simple thermostatically controlled heating system such as an electric iron, parallel logic can be used to determine when the temperature is below the set point and to switch on the heat. Using parallel logic to control the mode of the analog computer, iterative computations can be made for optimization studies, boundary value problems, and system identification studies. Similarly, preprogrammed parameter sweeps can be made using parallel logic to update the parameters of a system prior to a computer run, automatically make a computer run, and to repeat the process throughout, without active participation from the operator.

Parallel Logic

The term parallel logic implies a collection of logic components that can be wired together with patch cords to form independent logic networks. There are many components of each type and each performs its function at the same time—i.e., in parallel. This is in contrast with the traditional serial manner in which the digital computer carries out its calculations. Parallel operation enables logic circuits to perform many tasks such as keeping track of time, counting, decision making, memorizing past conditions all at once without serious time skew.

Logic Signals

At any given time a logic signal is either a logic ONE or a logic ZERO. This is similar to the case of an ordinary electric switch. It is either "on" or "off". Any other state different from "on" and "off" cannot exist and is not of interest to us. Several names exist which are frequently used interchangeably to describe the same logic level, such as

$$ONE = ON = HIGH = \text{``1''} = TRUE$$
$$ZERO = OFF = LOW = \text{``0''} = FALSE.$$

The Inverter

The inverter provides logical negation, a vital function needed in any logic program. Its symbol and truth table is shown in Fig. VIII-1. When the input is a logic ONE the output is a logic ZERO. When the input is a logic ZERO the output is a logic ONE. The output of the inverter is called the complement of the input. All components with a logic output contain an inverter so that both normal and complementary outputs are available for use. Therefore inverters do not appear as separate components but are prepackaged with other components. The inverter is depicted as a triangle with a line through it. The wires carrying logic signals to and from the inverter are shown as dashed lines to distinguish them from the wires carrying analog signals. If the input to the inverter is called "A" then using the usual convention the output is called "$\bar{A}$" (pronounced "not A" or sometimes "A bar").

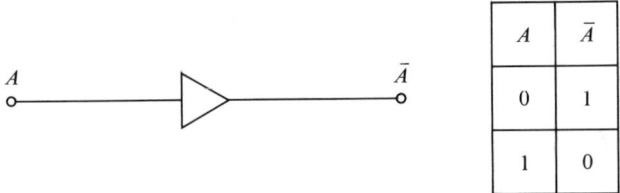

Fig. VIII-1. Programmer's Symbol and Truth Table for an Inverter

The AND Gate

The AND gate has two or more logic inputs. The output is a logic ONE if and only if *all* inputs are logic ONE. If any input is a logic ZERO the output is a logic ZERO. The symbol and truth table for the AND gate are shown in Fig. VIII-2. By convention, the AND operation can be written as

$$c = a \cdot b \text{ (pronounced a } AND \text{ b)}.$$

This operation is sometimes referred to as logical multiplication. Many of its properties do, indeed, resemble that of algebraic multiplication. Several references are given at the end of this paper for those who want to pursue Boolean algebra further.

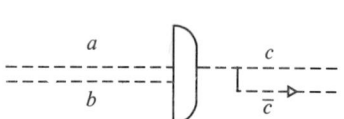

a	b	c
0	0	0
0	1	0
1	0	0
1	1	1

Fig. VIII-2. The AND Gate

The OR Gate

The OR gate has two or more logic inputs. If any input is a logic ONE the output is a logic ONE. The output is a logic ZERO if and only if *all* inputs are logic ZERO. The symbol and truth table for the OR gate are shown in Fig. VIII-3. The OR operation can be written as

$$q = m + n \text{ (pronounced m } OR \text{ n)}$$

This expression is often called logical addition for the same reason mentioned before.

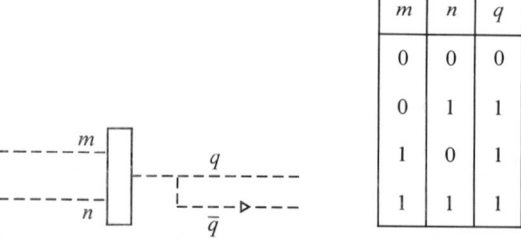

m	n	q
0	0	0
0	1	1
1	0	1
1	1	1

Fig. VIII-3. Programmer's Symbol and Truth Table for an OR Gate

Most logic systems do not contain OR gates. Therefore OR is obtained using AND gates and inverters and DeMorgans Theorem in the form $a + b = \overline{(\bar{a}) \cdot (\bar{b})}$. In other words, to obtain $a + b$ one has to perform the AND operation of their complements, $\bar{a}$ and $\bar{b}$, then take the NOT output of that operation. For those unfamiliar with Boolean algebra, the truth table in Fig. VIII-4 may offer an alternative convincing argument. Also shown on Fig. VIII-4 is the manner in which logical circuit is patched to obtain the OR operation. It should be clear to the reader now that to make one OR gate the operator needs only one AND gate.

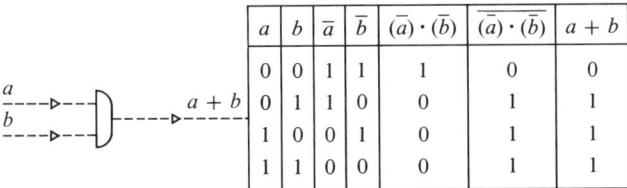

a	b	$\bar{a}$	$\bar{b}$	$(\bar{a}) \cdot (\bar{b})$	$\overline{(\bar{a}) \cdot (\bar{b})}$	$a + b$
0	0	1	1	1	0	0
0	1	1	0	0	1	1
1	0	0	1	0	1	1
1	1	0	0	0	1	1

Fig. VIII-4. Obtaining the OR Function Using AND Gate

Example 1: Draw a diagram to perform the following operation

$$Z = (A \cdot \bar{B} \cdot C) + \overline{(A \cdot C)}.$$

Figure VIII-5a shows the recommended programmer's diagram for obtaining Z and Fig. VIII-5b shows the diagram of the wiring. Notice that it is easier to reconstruct the original Boolean expression from Fig. VIII-5a.

(a) Programmer's Diagram   (b) Actual Wiring

Fig. VIII-5. Diagram for Example 1

Example 2: Find the values x and y such that the cost function C is minimum:

$$C = .20 + .12x + y - xy + 0.20y^2 - x^2$$

where $0 < x, y < 1.0$ and x and y cannot be found within the cross-hatched region shown in the following diagram.

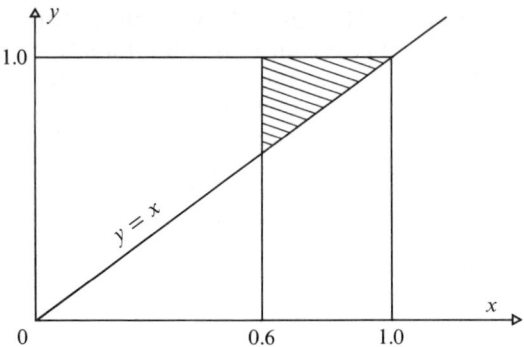

In Boolean language, the constraint can be represented by a logical signal which we name INHIBIT:

$$\text{INHIBIT} = (x > 0.6) \cdot (y > x).$$

The above equation can be interpreted as follows: INHIBIT will be at logical "1" if x is greater than 0.6 *AND* y is greater than x. Note that the expression $(x > 0.6)$ is a logical expression. It assumes the value "1" when $x > 0.6$ and a "0" when $x \leq 0.6$. How the $x - y$ plane is searched is not shown on Fig. VIII-6. It could be swept, TV-style or by two sinusoidal waveforms of different phase and frequency. The parameters x and y are fed into an analog circuit (contents not shown) which calcu-

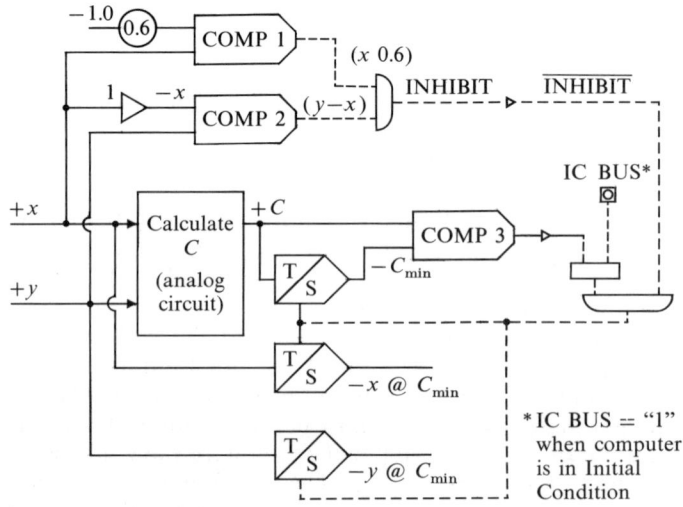

Fig. VIII-6. Circuit Diagram for Example 2

lates the cost function C. The three T/S amplifiers are to retain C_{min} and the coordinates (x, y) at which C is a minimum.

Initially the first value of C, x and y are tracked due to the "IC BUS" input to the OR gate. It is quite important to do so because the first search point may yield C_{min} and has to be retained. Once the computer goes to Operate mode the T/S amplifiers track only when a new minimum in C is found. If the search point falls into the restricted area the INHIBIT signal prevents the circuit from updating itself to the illegal minimum.

Flip-Flop (Bistable Multivibrator)

The flip-flop is a logic memory which can retain one bit of information indefinitely until disturbed by an external signal. By virtue of the above definition, each core of the digital computer memory is a flip-flop; so is the lamp switch which stays in the "on" condition until switched off. Symbolically the flip-flop is represented by the symbol on Fig. VIII-7. The inscriptions on the symbol denote:

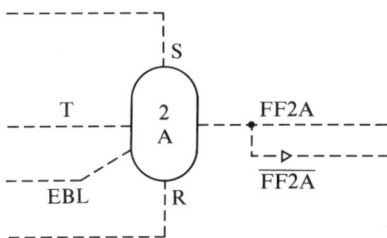

Fig. VIII-7. Programmer's Symbol of a Flip-Flop

2A = name of the flip-flop
EBL = Enable. When this input is low flip-flop 2A will not change state regardless of other inputs
FF2A = normal output of flip-flop 2A
$\overline{FF2A}$ = complementary output of flip-flop 2A
S = Set. An input of "1" to this terminal will cause the output FF2A to become "1" at the next clock pulse. By convention the clock pulses are not shown but it is assumed to be permanently connected to the flip-flop.
R = Reset. It forces the output FF2A to the "0" (or reset) state.
T = Trigger. The presence of a "1" at this terminal will cause the flip-flop to change its state at every clock pulse. Thus if a "1" is permanently patched into the T terminal of a flip-flop, the output of such a flip-flop will be a square wave whose frequency is half that of the clock. The triggering effect can also be achieved by simultaneously raising both the S and R terminals to the "1" level.

Figure VIII-8 sums up pictorially all the possible operations of a flip-flop. Note that all the clock pulses are numbered for easy reference in the discussion. Again the reader should bear in mind that the inputs into S, R, T and EBL come from other synchronous devices within the computer. As the result when they change their state they do so immediately *after* the appropriate clock pulse.

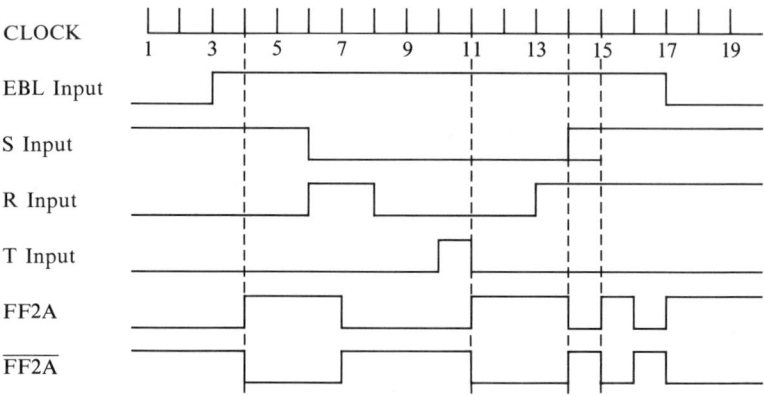

Fig. VIII-8. Flip-Flop Operation

In Fig. VIII-8 it is assumed that FF2A = "0" at the beginning. This output stays at zero up through clock pulse #4 because the EBL input is low. By convention the EBL input, as shown, is changing state *after* clock pulse #3. Therefore its enabling effect on the flip-flop won't take place until clock pulse #4. At clock pulse #11, FF2A is reset again because of the presence of the T input which instructs FF2A to change its state from a "1" to a "0". At this point it is convenient to learn one more definition, a *blip*. A blip is a positive going pulse that lasts for the period of the clock rate. The "T" input is a blip because it stays high for the period between clock pulses #10 and #11. Starting at clock pulse #15 the presence of both S and R inputs causes FF2A to trigger at every clock pulse, resulting in the familiar square-wave pattern. However, beyond clock pulse #18 no other changes occur because the EBL level has been dropped.

The Differentiator

The differentiator is a device that produces an output blip whenever its input changes from logic ZERO to logic ONE. Such an input from a synchronous device would rise on the trailing edge of a clock pulse. The output of the differentiator would also rise on the trailing edge of that clock pulse and would fall on the trailing edge of the next clock pulse. Figure VIII-9 shows the programmer's symbol and timing diagram for the differentiator. It is a useful device for detecting a change in a logical variable.

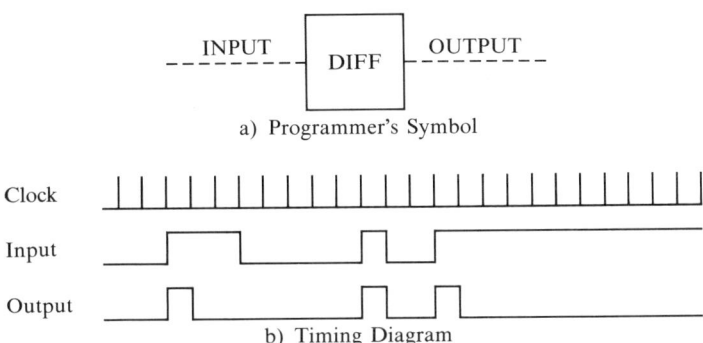

Fig. VIII-9. The Differentiator

Since the differentiator detects a ZERO to ONE step it is formally referred to as a leading edge differentiator. To detect a ONE to ZERO step the complement of the logic variable is used as input to the leading edge differentiator. Therefore trailing edge differentiators need not be supplied.

Example 3: Design a circuit which will monitor a variable x (initially $x > 0$) all the time and which will yield one blip whenever x becomes negative for the second time and thereafter (x is an analog variable).

A possible solution is shown in Fig. VIII-10. According to this figure as soon as the variable x drops below zero, $\overline{COMP}$ becomes high. This signal is differentiated by DIF which yields one blip and tries to set FF4C. By the time flip-flop 4C is set (after the next clock pulse) the blip out of DIF has gone by. But if the variable x becomes negative for the second time, $\overline{COMP}$ is differentiated as before and AND gate 5D will yield a blip because FF4C is already at the "1" level previously. From now on Fig. VIII-10 will give a blip every time x goes negative. Note that not all the inputs into flip-flop 4C are shown. As a rule whatever is not indicated on the diagram will not disturb the operation, i.e., the T and R inputs are implied to be "0" and the EBL input "1".

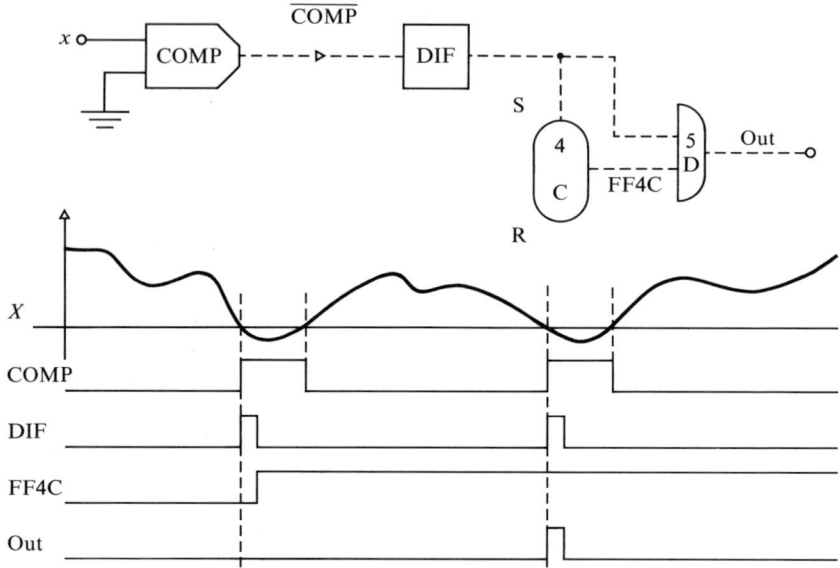

Fig. VIII-10. Diagram for Example 3

Example 4: Simulate the vertical motion of a bouncing ball.

The ball starts at an initial height (Y_0) and has an acceleration equal in magnitude to gravity but of opposite sign ($\ddot{Y} = -g$). When the ball hits the ground ($Y = 0$) its velocity ($\dot{Y}$) changes from a downward direction to an upward direction and its magnitude is attenuated by a factor K, ($\dot{Y}_2 = -K\dot{Y}_1$), where K is equal to the square root of the coefficient of restitution ($K = CR$). Figure VIII-11 shows the simulation diagram.

From the diagram it is seen that A10 and A11 have opposite mode of operation. At any time only *one* of them is in the Operate mode. Suppose A11 is in the IC mode and A10 is in Operate mode. They stay so until Y reaches zero at which time $\overline{\text{COMP}}$ becomes a ONE. This positive going level is differentiated and is used to trigger flip-flop 3C. Flip-flop 3C in turn forces A11 into the Operate mode and A10 into the IC mode. At the time of switching:

$$(\text{Output of A11}) = -\sqrt{CR} \cdot (\text{Output A10}).$$

Flip-flop 3C also turns on switch 2B and turns off 2A. When Y reaches zero again the flip-flop is triggered back to its original state and the integrators and switches go back to their original mode. The cycle continues until the ball loses its energy and no longer bounces. Since no provision has been made for this case, the simulation no longer applies.

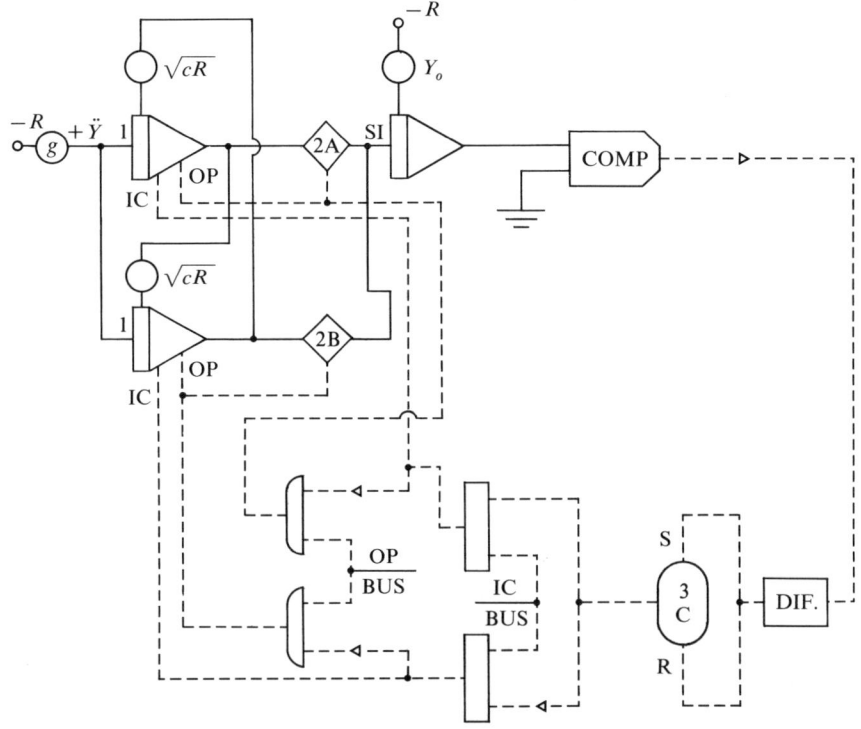

Fig. VIII-11. Bouncing Ball Diagram

The Monostable Timer

The monostable timer (mono) serves the same purpose as a toaster timer, i.e., once activated, it runs for a period of time then shuts itself off automatically. The mono output is normally low unless disturbed by a positive going input signal. It is only then that the output becomes high at the next clock pulse, remains high for a predetermined time, then drops back to the normally low level. If the input is constantly high the output will become high for a period, low for one clock period, high for a period, low for one clock period, etc. When the mono is high, input signals are ignored. The period that the mono remains high can be adjusted by the programmer from about 10 msec. to 10 seconds. Figure VIII-12 shows the programmer's symbol and timing diagram for the monostable.

The mono described herein is of a 100% duty-cycle type since it is able to set again, on command, one clock pulse after it was reset (clock pulse #22). Older monos of the 50% duty-cycle type required a recharging period (between "on" times) approximately equal to the "on" period. Of

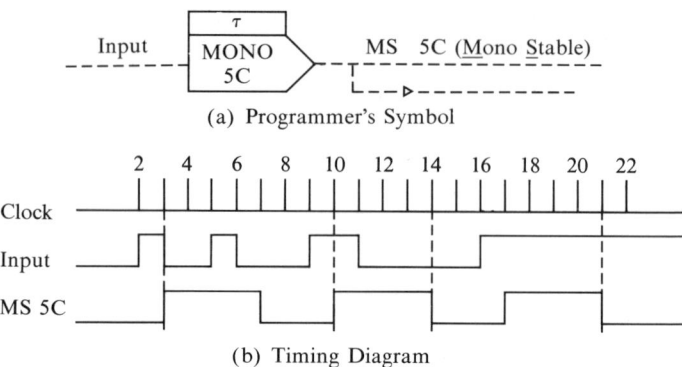

Fig. VIII-12. The Monostable Timer

interest is the shaded blip that occurs between clock pulse #5 and #6. This blip is entirely ignored because it arrives when the mono is still set. The period can be adjusted to an accuracy of a few per cent. Any applications requiring great accuracy should be programmed using counters which will be discussed later. However one should not have false ideas about the mono's usefulness. For instance, it is not desired to plot the entire run but only to store one value generated during some run that meets a certain criterion. With the computer in a repetitive rate of 10 solutions/sec., it is impossible to plot any value when the program decides to do so. A mono could be used to force the computer in IC mode for a period of 0.2 sec. whenever a decision to plot is reached. This time interval is intended to wait for the plotter to drop then lift its pen. Of course, there is no need to have an accurate timer in this case.

Registers

Register is the name given to a group of flip-flops internally connected and operating as a single unit. There exist two general types of registers classified according to their use: *Shift Registers* and *Counters*. In modern computers, general purpose registers are provided. The operator can make them counters or shift-registers at his own discretion by actuating a group of control signals.

Shift Register

The shift register is usually composed of 4 packaged flip-flops which respond to a set of control signals as illustrated in Fig. VIII-13. The output of each individual flip-flops A, B, C and D, their complements and their inputs R and S are accessible. All these terminals have been explained previously. The meaning of other terminals is as follows:

CLR = Clear, i.e., resets all 4 flip-flops
  L = Load, i.e., forcing each flip-flop to the state dictated by its S and R inputs
  SI = Serial Input. There are two methods of getting information into the shift-register (SR): by loading using the L input *or* by shifting it in via the SI input
  SO = Serial Output. Again to load the content of the SR, one can either take it out from each individual flip-flop outputs *or* obtain it at SO one bit at a time (serially). By connecting SO of one SR to SI of the next one can make SR with more than 4 bits
  SH = Shift. Information is passing from one flip-flop to the next, i.e., the serial input SI is transferred to FFD (flip-flop D) and that of FFD is passed over to FFC and so on. As long as SH = 1 the shifting continues at clock rate.

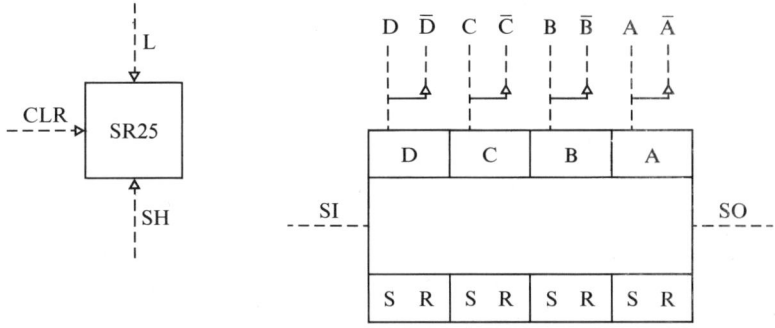

Fig. VIII-13. The Shift Register

For example, let the content of the register be at 1011 and the SH input is raised high for two consecutive clock pulses. The shift register will behave as follows:

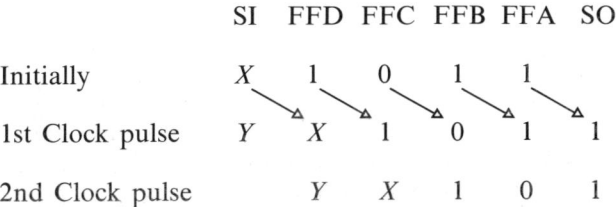

The arrows indicate how information propagates internally. The logical variables X and Y are status of the input patched into SI evaluated initially and at clock pulse #1 respectively.

Example 5: Consider Fig. VIII-14. This circuit consists of more than one $\overline{SR}$ cascaded end-to-end. The last two stages, $N - 1$ and N, are fed to a serial adder (half adder) whose output leads back to SI. An inspection of the truth table on the right side reveals the nature of the half-adder. If N is added to $N - 1$ and the carry is ignored the result would be as indicated by the 3rd column of this truth table. The half-adder is sometimes called EXCLUSIVE OR because of its similarity to the OR operation. Only one difference exists: when both N and $N - 1$ are at logic "1" they exclude each other, resulting in a "0" for output.

N	$N - 1$	Half-Adder
0	0	0
0	1	1
1	0	1
1	1	0

One should focus his attention on Fig. VIII-14 again. Starting with an arbitrary pattern of bits in the SR's (except all zeros) a pseudo-random pulse train of ONE's and ZERO's will appear at the serial output, SO, whenever the input signal, SHIFT, is high. It has been proved in literature that the resultant pulse train repeats itself after $2^n - 1$ shiftings. One should not be alarmed, because the programmer can always make N (number of flip-flops in the SR's) very large just by cascading more shift registers. This is a rather simple operation. A larger value of N will produce a lesser chance for a repetition. Naturally, a repetition will destroy the random characteristic of the pulse train. For the sake of illustration, suppose the shift register has 6 flip-flops ($N = 6$) and that initially the flip-flops assume this configuration: 101110. Then one can monitor SO and see the following pulse trains:

Counters

Unlike the shift-register the content of counters are interpreted as numbers. Most modern computers provide versatile counters that can count in binary or BCD. They can increment as well as decrement the count; all function can be controlled from the patch panel. Counter configurations vary widely from computer to computer; only general properties are discussed here. One should be aware of this fact when using

011101100110101011111100000100001100010100111101000111001001
011011101100110

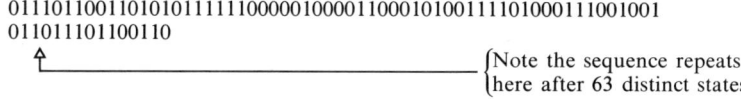

{Note the sequence repeats here after 63 distinct states

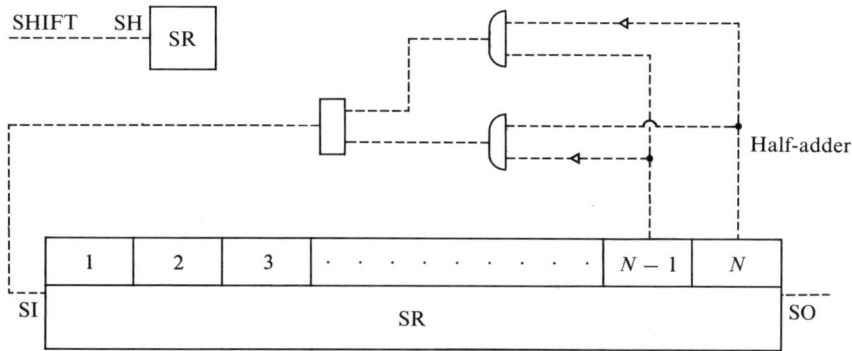

Fig. VIII-14. Pseudo-Random Number Generator

a counter in an unfamiliar computer because this counter may not have two or more of the features discussed below.

Figure VIII-15 depicts the symbol for a four-bit counter (CTR). The maximum count contained in this counter is 15. By cascading more than one CTR end-to-end a larger number could be attained. All the terminals described previously (CLR, L, A, $\overline{A}$, etc.) will not be repeated here. The significance of the others are:

BCD = Binary Coded Decimal. When this input is raised, the counter will count in the decimal system (radix 10).

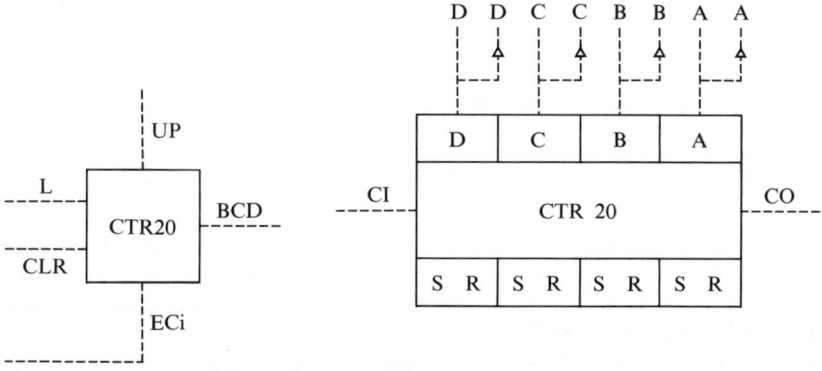

Fig. VIII-15. General-Purpose Counter

The following sequence will explain the concept:

	FFD	FFC	FFB	FFA	CO
Initially	0	1	1	1	
Clock pulse #1	1	0	0	0	
Clock pulse #2	1	0	0	1	
Clock pulse #3	0	0	0	0	1

Initially the counter content is 7. Each subsequent clock pulse increments it to 8 then 9. But after the count of 9 one more count resets all flip-flops.

If BCD were low instead, the counter would keep incrementing past 9, until 15. It is only then that it resets itself.

UP = As its name implies, this input causes the count to increase. The absence of this signal (=0) makes the CTR to count down. This feature finds important applications such as in the rate resolver. In such a resolver only the angular velocity $\dot{\theta}$ is given and the number of complete revolutions (maybe forward or backward) is kept track of using the counter.

If $\dot{\theta} > 0$ and $\theta = 2\pi \longrightarrow$ increment the count
If $\dot{\theta} < 0$ and $\theta = 2\pi \longrightarrow$ decrease the count

CO = Carry Out. It also means BORROW in case of down counting.
CI = Carry In. Usually this signal comes in as a blip resulting in one count. If CI is at level ONE the CTR will count at clock rate.
ECi = Enable Carry In. A ZERO into ECi will prevent CTR from counting regardless of the state of CI.

Example 6: Calculate, on line, the mean-square of an analog signal $f(t)$ read back from an FM tape. Assuming that the data lasts for 1000 seconds and the maximum value of the mean square $\overline{f^2(t)}$ varies greatly from run to run. One method of approximating $\overline{f^2(t)}$ is:

$$\overline{f^2(t)} \equiv \frac{1}{T}\int_0^T [f(t)]^2 \, dt. \qquad (1)$$

Chances are that the integration of $f^2(t)$ over 1000 seconds will be a very large number. By conservative analog scaling technique the coefficient of P10 will be so small that appreciate error will result. Figure VIII-16a offers a scheme that will extend the dynamic range of the integrator A30 almost without limit. The idea is to set P10 as high as one wishes to assure good accuracy. Suppose that FF1D starts out to be ZERO. It in turn causes the input into A30 to be negative. As A30 integrates, its output reaches +Reference voltage and it is detected by the comparator COMP. Every time that COMP changes state FF1D is triggered due to the leading-edge and trailing-edge differentiators combination. Once FF1D = 1, a positive $+f^2(t)$ is fed to A30 forcing its output to decrease.

Circuit diagram

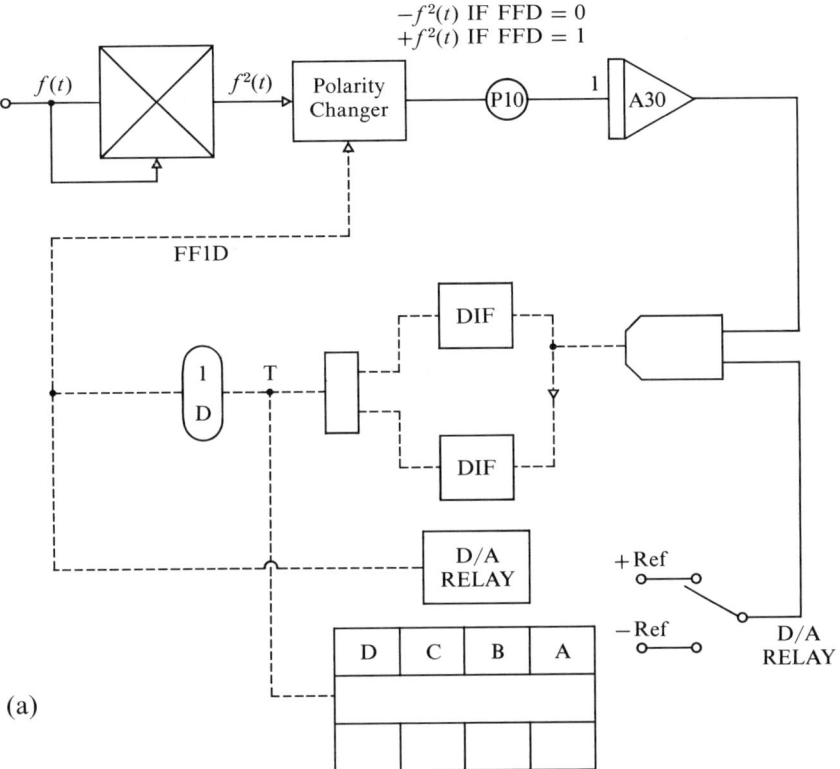

(a)

Timing diagram

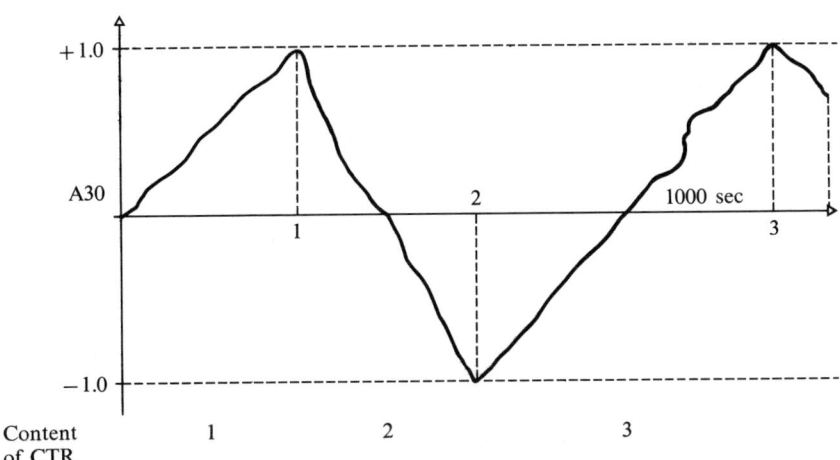

Fig. VIII-16. Estimation of Mean-Square Values

A representative waveform of A30 is drawn in Fig. VIII-16b. Let the computer run be stopped as indicated in Fig. VIII-16b. At that time the content of CTR is 3 and

$$\int_0^{1000} [f(t)]^2 \, dt = 5.5 \text{ or } 550 \text{ volts in a } \pm 100\text{V.}$$
Reference System

Example 7: One form of the Mathieu's equation is written as

$$\frac{d^2y(t)}{dt^2} + (a - 2q \cos 2t) \cdot y(t) = 0. \tag{2}$$

This classical equation not only describes vibrating strings and membranes but also waveguides, FM circuits and sloshing phenomena inside a fuel tank. It is an interesting problem to solve because it tends to be unstable for a certain combination of "a" and "q". Using parallel logic, write a program that will automatically perform the following operations:

(1) Solve the differential equation repetitively.

(2) Investigates all possible combinations of the parameters "a" and "q" where

$$0 \leq a \leq 12.5$$
$$0 \leq q \leq 5.0.$$

In the time domain, parameters "a" and "q" are searched as shown below.

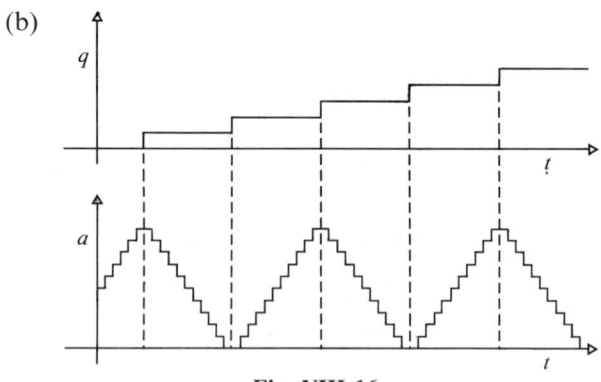

Fig. VIII-16

If one sweeps parameter "a" in 100 steps and "q" also in 100 steps then the Mathieu's equation must be solved 10,000 times in order to cover the whole a-q plane.

(3) The variables "a" and "q" drive the X and Y of the plotter respec-

tively. Whenever the solution $y(t)$ proves to be stable, the program will order the pen to drop, otherwise it is lifted. An arbitrary definition of stability is that $-1.0\ y(t) +1.0$ for a period of 20 msec. computer time (or 20 sec. real time).

(4) Terminates the run after 10,000 solutions. This is accomplished by monitoring the value of "q".

At this point the flow-chart is needed by the programmer to plan his overall strategy before he actually draws a circuit diagram. Figure VIII-17 resembles a digital program flow-chart because it utilizes the same block symbols, but that is where the similarity is ended. Its multiple branchings from the same point characterizes *parallel* logic because it makes parallel decisions and is able to carry out many tasks at the same time.

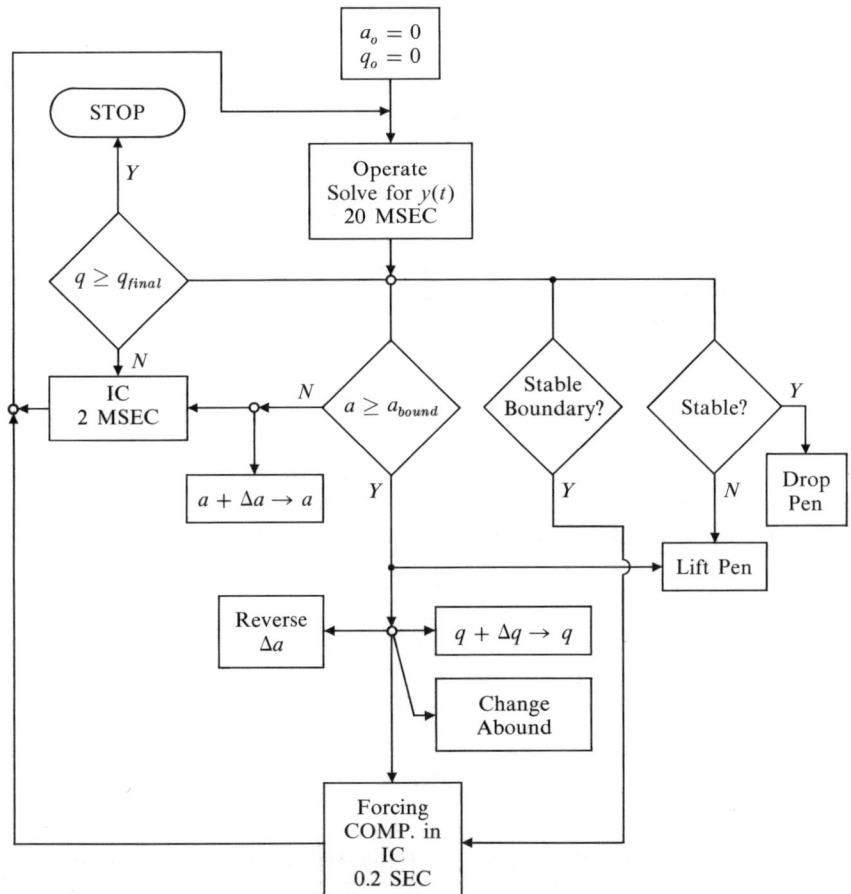

Fig. VIII-17. Flow Chart for Example 6

Normally the repetitive operation (rep-op) cycle is divided into 2 msec. and 20 msec. for the IC and OP modes respectively. However, the IC period is lengthened to 200 msec. to wait for the plotter pen to lift in case of an excursion from the stable to the unstable region or pen dropping in the reverse situation. Similarly it is necessary to lift the pen and wait for 200 msec. when parameter "a" reaches its extrema at which points parameter "q" is updated. The reason for this delay is to allow the pen to go from "q" to "$q + q$", because it cannot cover this distance in 2 msec. normally allocated for the IC period.

When drawing the diagram on Fig. VIII-18a and b an attempt was made to avoid hardware-dependent features. Therefore it should be emphatically clear that in order to make this program work on any analog computer, a slight modification is needed to make the controls compatible. The succeeding are a few salient points of the program:

(1) There exist two accumulators, one for sweeping the "a" parameter and the other, "q" parameter. The former updates itself once every Operate cycle and the latter, once every 100 cycles.

(2) To lessen the chance of confusion consider the TIMER as a blackbox having one input and two output terminals (Fig. VIII-18b). Under normal conditions the "OP BUS" cycles between "1" for 20 msec. and "0" for 2 msec. The timing intervals are usually controlled by thumb-switches which can be set by the programmer. Once started, this timer pulses continuously and is interrupted only by the input (IC). As long as the (IC) input remains at "1", it clamps the "OP BUS" to ZERO level, holding the computer in Initial Condition mode. "IC BUS" is the complement of "OP BUS".

(3) "UPDATE Q", a logical variable, goes high whenever 'COMP 21' changes its state. Comparator 21 also controls the sign by switching + Ref or − Ref into A47 (Fig. VIII-18b).

(4) Unstable solutions are detected by COMP 38 which, in turn, sets flip-flop 6B. Thus FF6B "remembers" the fact that $y(t)$ overloads sometimes during one Operate cycle and passes this information to FF7A at the completion of this cycle. It is FF7A that controls the plotter pen and, by pulsing MS6, makes the TIMER wait.

A representative plot by the program is shown in Fig. VIII-19. The shaded area coincides with the stable region and the general contour checks accurately with theoretical prediction.

Conclusion

This chapter intended to indicate the various types of logic components and how one may go about programming using these components.

At the conclusion of this chapter, one should have an adequate understanding of the basic logic components and how they are utilized in hybrid computation.

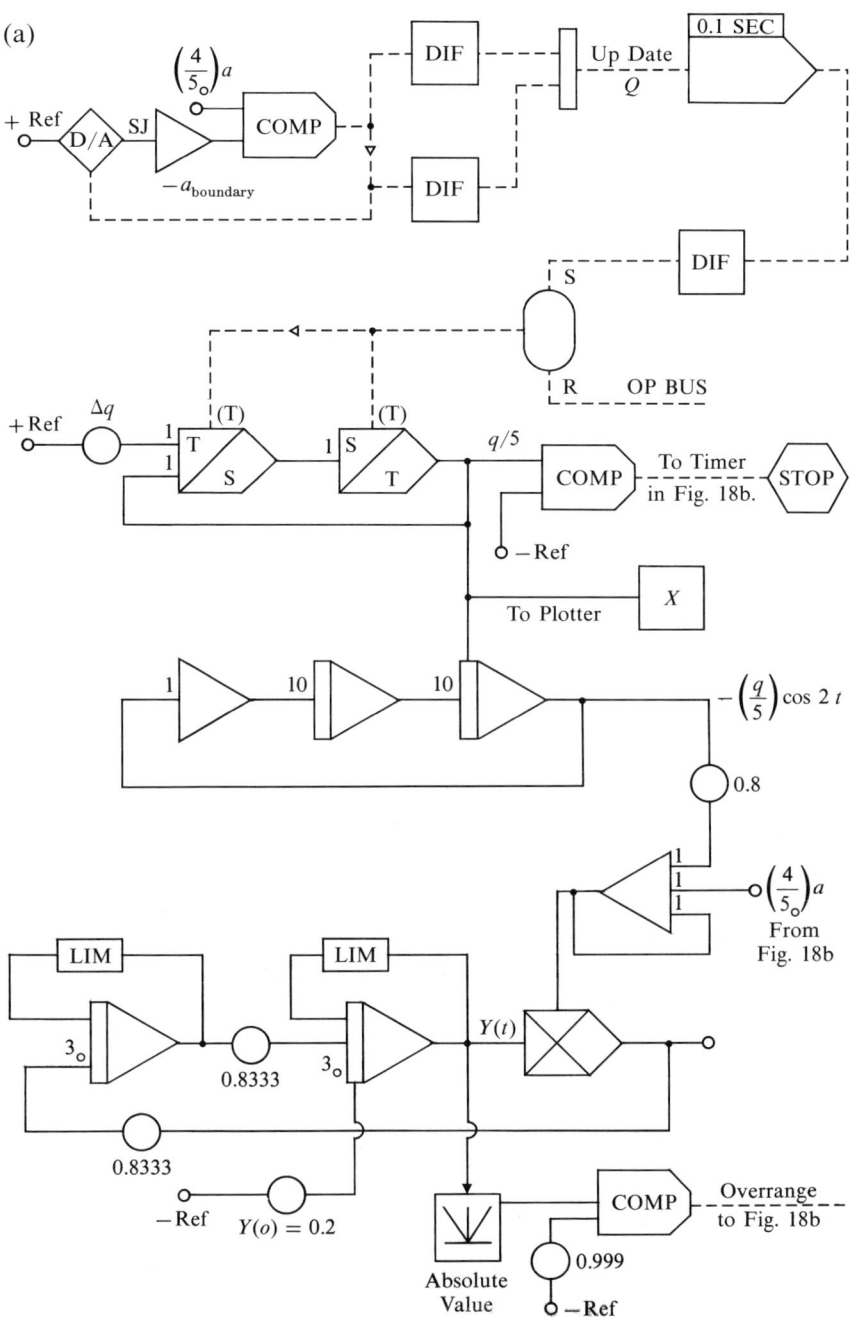

Fig. VIII-18. Circuit Diagram for Example 6

(b)

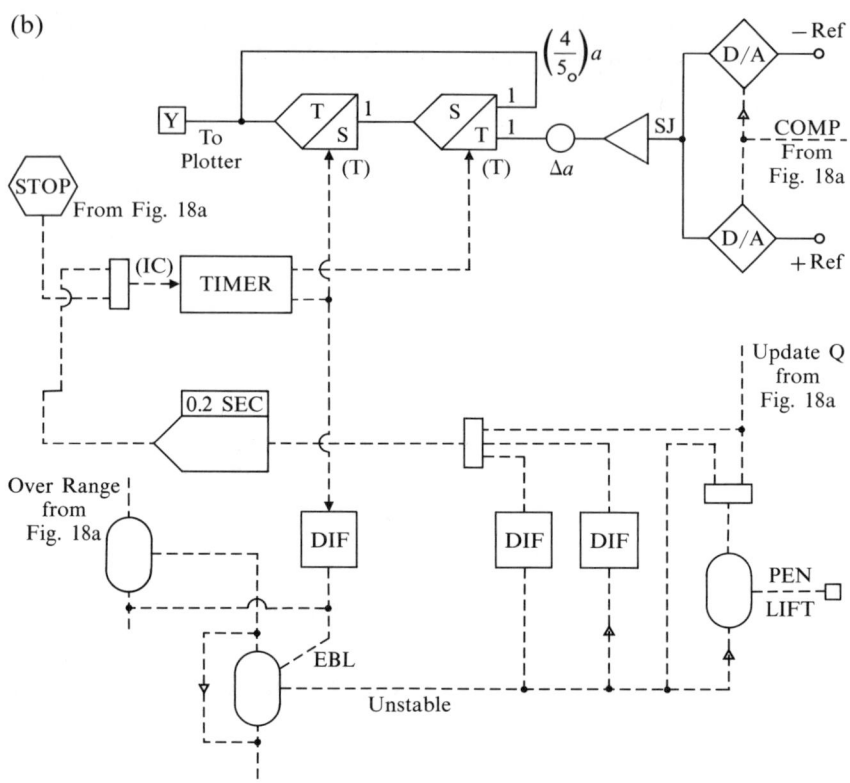

Fig. VIII-18. Circuit Diagram for Example 6

Since this is the last chapter of the book it should be mentioned that an attempt has been made in the preceding eight chapters to provide the reader with the information necessary to do Analog/Logic computer programming and simulation. It is up to the reader to extend this knowledge into use by actually getting on a computer and solving problems. Good Luck!

Homework & Exercises

Problem 1: Prove that the circuit depicted in Fig. VIII-20 is a leading-edge and trailing-edge differentiator, i.e., it will yield a blip every time that X changes its state.

Problem 2: Design a circuit that tracks an analog variable $y(t)$ and retains

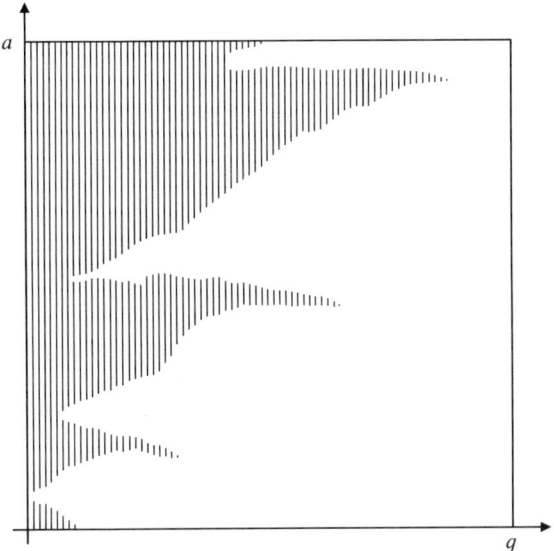

Fig. VIII-19. Stability Plot for Mathieu's Equation

the maximum and minimum points. The extrema may or may not have the same sign.

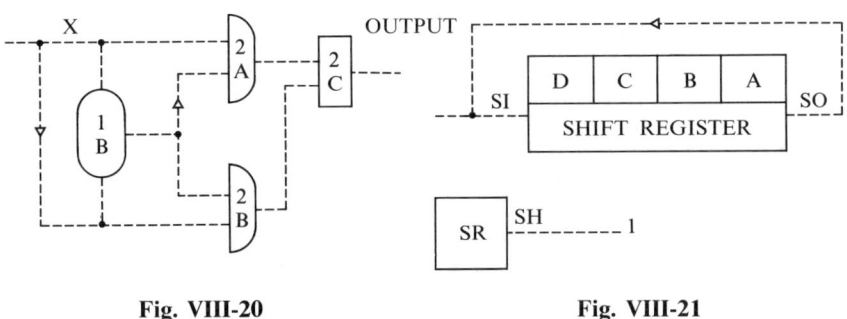

Fig. VIII-20   Fig. VIII-21

Problem 3: Given the steady-state heat conduction and radiation equation in one-dimensional space:

$$\frac{d^2T}{dx^2} = KT^4$$

where T = temperature
 x = independent variable
 K = a constant.

The boundary conditions are given as

$$T(0) = T_o$$
$$\frac{dT(L)}{dx} = 0.$$

Write an iterative program that will satisfy the second boundary condition by correcting the trial value $dT(0)/dx$.

Problem 4: Fig. VIII-21 describes a "TRANSPOSED SHIFT REGISTER". Assume that all flip-flops are ZERO initially, show that there can be only 8 distinct combinations as the shifting is being performed. Also indicate that only 8 two-input AND gates are necessary to decode the 8 states of the SR. (It should be recognized that a 4-bit SR can hold 16 distinct combinations ranging from 0000 to 1111 binary, but in this instance some combinations can never be formed).

Problem 5: Hybrid programs often require a synchronous pulse train of some frequency f. Verify that, in Fig. VIII-22, "A" is a square-wave, "B" an analog triangular-wave and "P" a pulse train. All three signals are of the same frequency f. Derive the expression of f as a function of K, the pot setting of $P1$, and the time-constant of the integrator A20.

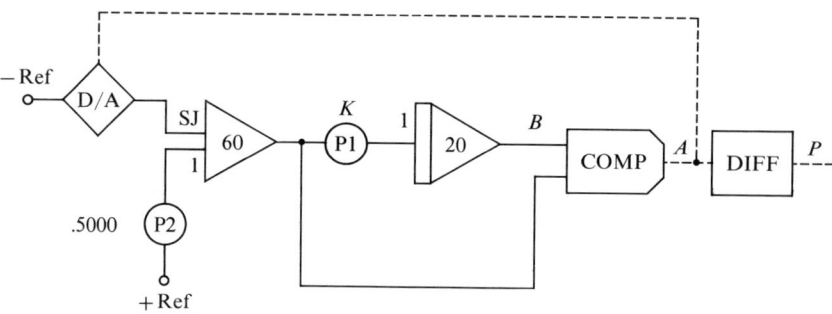

Fig. VIII-22

Problem 6: Draw a circuit diagram that generates the same pulse train P as in problem 5 without using analog components. (Hint: One should make use of counters to keep track of the time intervals between pulses). On the computer, patch problems 5 and 6 and demonstrate that both signals P can be superimposed perfectly on each other.

Problem 7: Let $N(x)$ be a stationary random variable. The probability that $N(x)$ lies in the "infinitesimal" interval x and $x + \Delta x$ is defined to be $f(x)$. In this example let $N(x)$ be the output of a low-frequency Gaussian noise generator; its density function can be estimated using the scheme illustrated in Fig. VIII-23a. Patch and run this program 100 seconds each time. Start at $x = 0$ and $x + \Delta x = .01$ or 1% of reference voltage; increment x to $x = .01$ and $x + \Delta x = .02$ then $x = .02$ and $x + \Delta x = .03$ and so on. Each run results in a single value $f(x_i)$. By plotting $f(x_i)$ versus x_i, one can obtain a histogram of the noise generator (Fig. VIII-23b). The circuit contained in Fig. VIII-23a works only for positive values of x_i. It is unnecessary to calculate the other half of the curve because of its symmetry about the $x_i = 0$ axis.

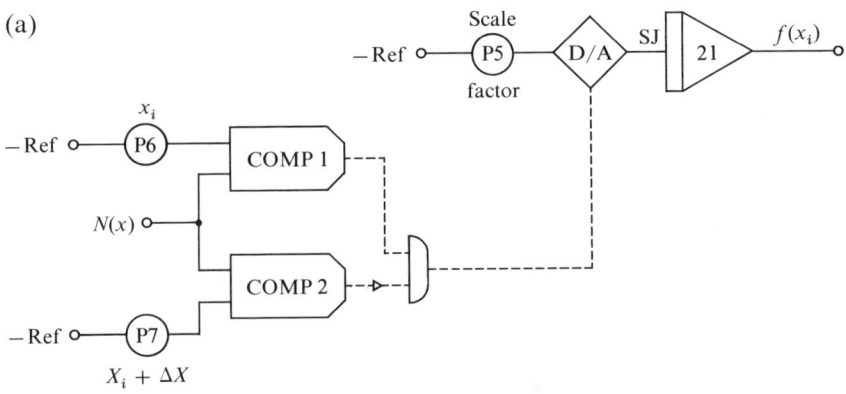

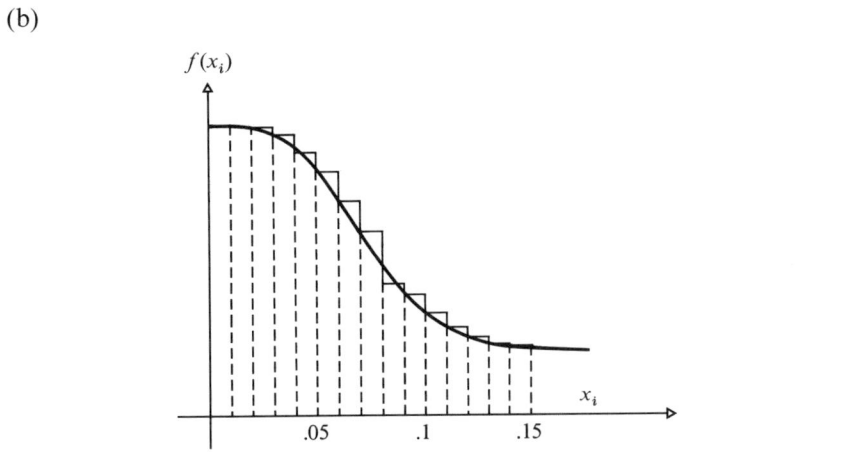

Fig. VIII-23

References

1. R. Bennet, and A. Fulton, "The Generation & Measurement of Ultra Low Frequency Random Noise." Project Cyclone Symposium 1, March 1953.
2. R. S. Berkowitz, *Modern Radar Analysis & System Design* (New York: John Wiley, 1965).
3. S. Caldwell, *Switching Circuits and Logical Design* (New York: John Wiley, 1960).
4. Alan Carlson, "Education & Training Memo No. 20," Electronic Associates, Inc., Princeton, N.J., 1963.
5. Electronic Associates, Inc., Princeton, N.J. "Hybrid Computation," 1964.
6. G. Hannauer, "Basics of Parallel Hybrid Computers," Electronic Associates, Inc., Princeton, N.J.
7. Hannauer, Serlin, Holsberg, "Automatic Iterative Operation on an Analog Computer," EAI Applications Reference Library.
8. G. A. Korn, and H. Huskey, *Computer Handbook* (New York: McGraw-Hill Book Co., 1962).
9. P. Landauer, "Spectrum of Applications for the Modern Hybrid Computer," EAI Applications Reference Library.
10. Omri Serlin, "Automatic '3-D' Plotting with the EAI TR-48/DES-30 Desk Top Analog/Hybrid Computing System," EAI Applications Reference Library.
11. N. Zierler, "Linear Recurring Sequences," *Jour. Soc. Industr. & Appl. Math.* 7: March 1959.
12. N. Zierler, "Several Binary Sequence Generators," MIT Technical Report No. 95, Sept. 12, 1965.

APPENDICES OF ANALOG COMPUTER PROBLEMS

Introduction to Appendices

The Appendices A-J are intended to give the student a thorough foundation in the solution of analog computer problems. The problems were intentionally drawn from many disciplines to give the student the widest possible selection while at the same time introducing him to engineering concepts in many fields.

Each Appendix considers a particular problem and can therefore be considered alone. The author has found that a very convenient technique for considering the problem is to hand out the statement of the problem and the sheet Problem Statement I and then ask the students to do that portion of the problem. When that is completed, the solution to Part 1 is handed out. This procedure is continued for Problem Statement II and III until all sections have been completed.

It is hoped that these Appendices will be an aid to teachers and to students in actually learning to program an analog computer.

APPENDIX A

Spring Mass Damper System

Introduction

The equation below represents the motion of a linear spring mass damper system; the computer diagram represents a simulation of the system. This exercise is designed to introduce one to patching and running an analog computer. Refer to any available reference handbooks or notes given you for patching instruction, but don't hesitate to ask your instructor for aid.

Problem Statement

for
$$m\ddot{x} + d\dot{x} + kx = 0$$

$$m = 0.699 \text{ slugs}$$

$$d = 0.300 \text{ lb./ft./sec.}$$

$$k = .500 \text{ lb./ft.}$$

and $\quad x = 10$ ft., $\dot{x} = 0$ ft./sec. at $t = 0$.

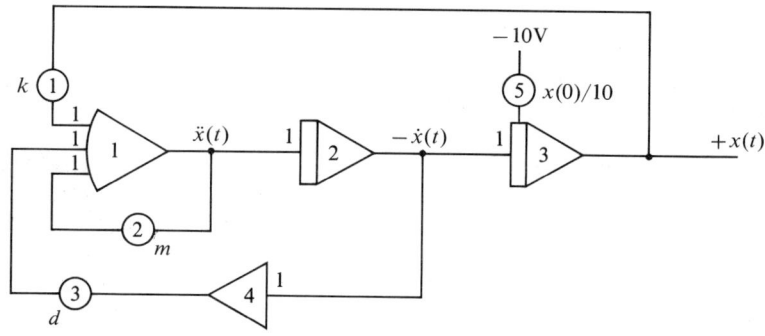

Fig. A1

1. Fill in the following table

Pot No.	Parameter	Setting
1	k	
2	m	
3	d	
5	$x(0)/10$	
15	?	

2. In general, we wish to plot x, $\dot{x}$, and $\ddot{x}$ versus time. Actually, we will plot the voltages proportional to x, $\dot{x}$, and $\ddot{x}$ versus a voltage which is proportional to time, i.e., $y = at$. (The function y is often called a "ramp.") The following circuit is proposed as a ramp generator.

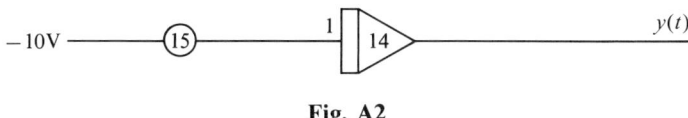

Fig. A2

If we want $y = +10V$ when $t = 20$ sec., what should the setting be on pot 15? Add this to the table.
3. Patch the program shown in the computer diagram, and the ramp generator from part 2. Mark off each connection on the diagram with a colored pencil as the connection is made. Always work from the input of a device to the output of the element which "feeds" that device. For example, one patching connection might be described as "patch from the input of pot 2 to the output of amplifier 1."
4. Insert the patch panel into the computer and set all pots to their correct numerical settings. There should be *no* overload light lit! If in doubt, check your reference material and/or instructor. Remove the connection from pot 1 to amplifier 1 and check the value of the setting of pot 1. Are the settings different? Why? Be sure to repatch that connection.
5. Patch the output of A14 into the arm input (X) of the X-Y plotter, and the output of the selector into the pen input (Y). Check to see if the plotter inputs are terminated on the readout module of the computer. Adjust the plotter so that a plot of x vs. y fills the paper vertically and so that 20 seconds of running allows most of the horizontal axis of the paper to be traversed. With these plotter settings, plot x, $\dot{x}$, $\ddot{x}$ vs. t on the same sheet of paper. Label the graph (see results).
6. Using a fresh sheet of paper, change the pen setting so that x vs. t, $\dot{x}$ vs. t, $\ddot{x}$ vs. t can be plotted on parallel tracks without overlapping. Label appropriately. Now, double the value of d, and make the same runs on the same parallel tracks. (If possible, your instructor will supply you with a contrasting colored ink for the plotter pen before making the three runs with the doubled value of d.)

Automobile Suspension System

Introduction

The simulation of physical systems on the analog computer is one of its most powerful functions. By simulation techniques, one is saved the costly situation of building a system and changing the components for

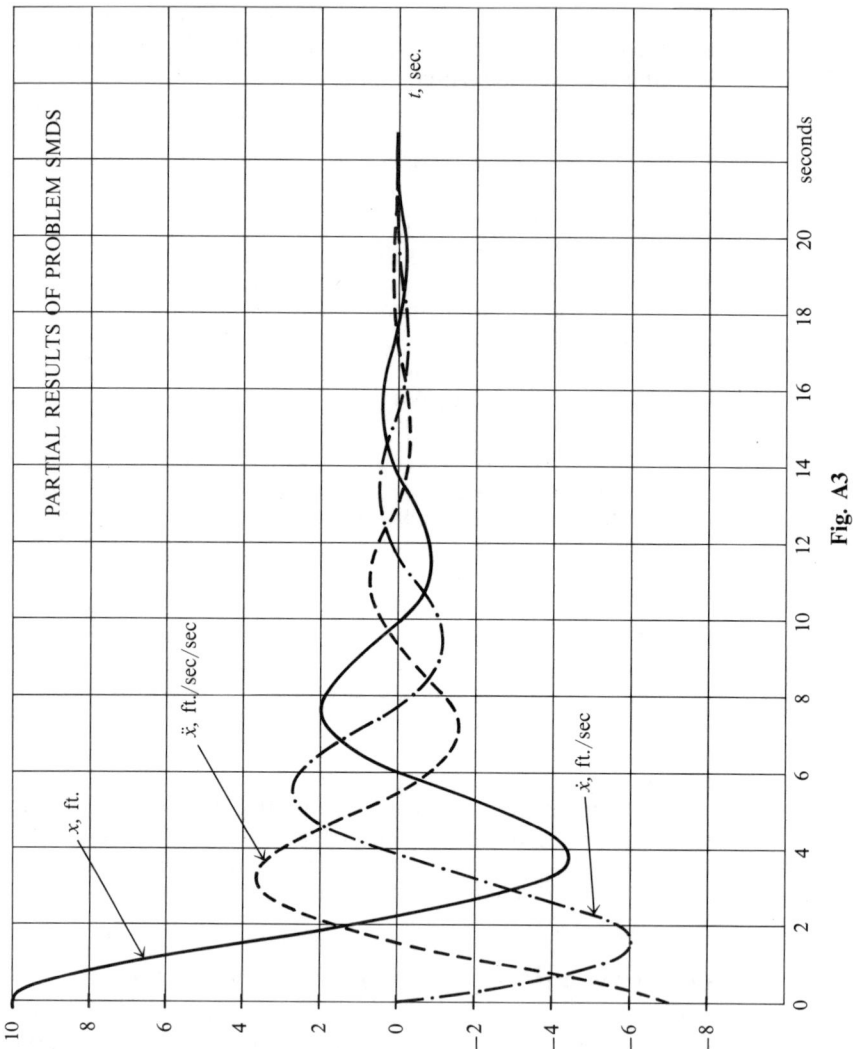

Fig. A3

a parameter study. Simulation on the computer enables one to study a system, with all of its variations, without physically building it.

Coupled mechanical systems usually require careful study and experimentation to guarantee their correct behavior. This problem presents a particular type of coupled system, namely a simplified version of an automobile suspension system, for simulation on the computer. Although there are many variations of this system, the one to be studied will have all of the component values specified.

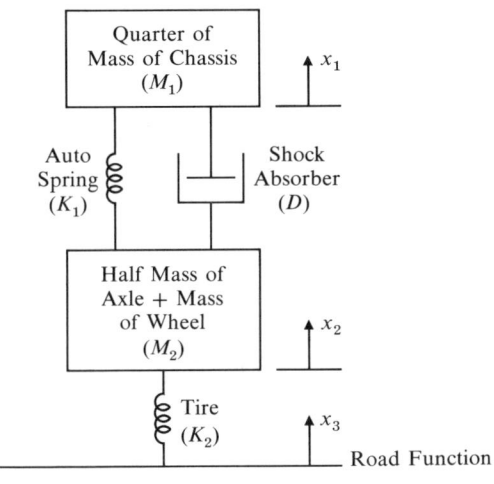

Fig. A4

PROBLEM STATEMENT I

Consider the system in Fig. A4, which is a simplified model of one wheel of an automobile suspension system. The spring action of the tire provides us with one equation and the action of the auto spring and shock absorber another. A force-balance yields the eqs. (A1) and (A2):

$$M_1\ddot{x}_1 + D(\dot{x}_1 - \dot{x}_2) + K_1(x_1 - x_2) = 0 \quad (A1)$$
$$M_2\ddot{x}_2 + D(\dot{x}_2 - \dot{x}_1) + K_1(x_2 - x_1) + K_2(x_2 - x_3) = 0. \quad (A2)$$

For simplicity, we may assume that x_3 is a step function (the car might be riding up onto a curb). Other, more complicated functions are possible, of course; however, a step input keeps the program simple. The quantities K_1, K_2, M_1, M_2, D, and x_3 are all constant for a given computer run. These equations should be programmed so that exactly eight (8) amplifiers are utilized. Hint: generate terms involving the velocity differences and displacement differences. The values of the parameters encountered are

$$M_1 = 25 \text{ slugs}$$
$$M_2 = 2 \text{ slugs}$$
$$x_3 = 5 \text{ inches} = 5/12 \text{ ft.}$$
$$20 \leq D \leq 200 \text{ lb./ft./sec.}$$
$$K_1 = 1000 \text{ lb./ft.}$$
$$K_2 = 5000 \text{ lb./ft.}$$

This problem has been considered thoroughly in the text in Chapters III and IV and discussion of it will not be continued here.

APPENDIX B

Tubular Chemical Reactor Control System

IN CHEMICAL CONTROL SYSTEMS as well as all types of controlled systems, the designer is faced with the problem of stability. This problem illustrates the analysis of a reactor control system as to how stability depends on the proportional gain of the controller. The system to be considered is illustrated in Fig. B1.

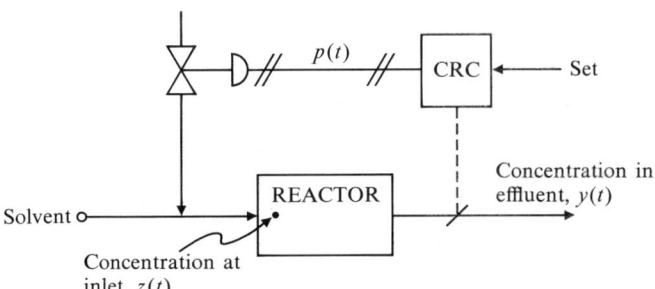

Fig. B1. Flow Diagram

Under laminar flow conditions, the response of a tubular mixing section can be approximated by the differential equation

$$T^3\dddot{y} + 3T^2\ddot{y} + 3T\dot{y} + y = z. \tag{B1}$$

For the sake of simplicity, assume that the manipulating flowrate x is related to the concentration z by the constant $K(\text{mole/ft.}^3)/(\text{ft.}^3/\text{min})$. The control system consists of:

(a) a fast-acting transducer which monitors y and reports it as an electrical signal $K_T y$ to the CRC.

(b) electro-pneumatic controller having a gain K_c, the output $p(t)$ of which controls the CRC.
(c) fast-acting linear valve which manipulates the trace flowrate, $x = K_v p$.

The entire system, shown in Fig. B2, is described by
$$\dddot{y} + (3/T)\ddot{y} + (3/T^2)\dot{y} + (1/T^3 + aK_c K_T)y = aK_c r(t) \tag{B2}$$
$$a = K_v K/T^3$$

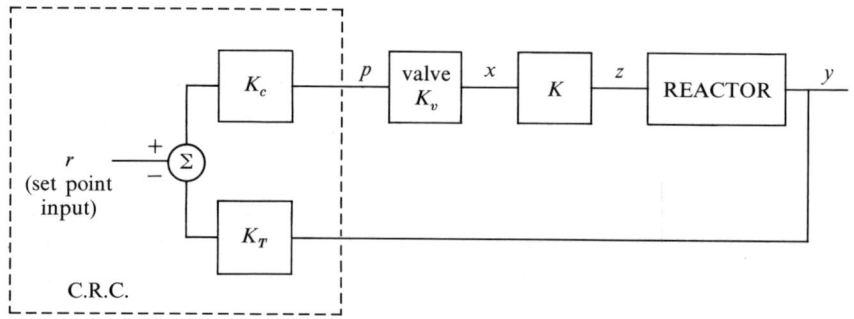

Fig. B2. Block Diagram

where $r(t)$ is the controller set point input
and K_c is the gain of the controller portion of the CRC, $1 \leq K_c \leq 27$ psi/mv
         K_T is the gain of the transducer portion of the CRC, 1000 mv/(mole/ft.^{3})
         K_v is the valve "manipulation constant," 0.100 (ft.^{3}/min.)/psi
         K is the ratio of the manipulation flowrate to the concentration z, 0.010 (mole/ft.^{3})/(ft.^{3}/min.)
         T is the reactor time constant, 0.500 min.
Thus, $a = 0.0080$ (mole/ft.^{3})/(psi)(min.^{3}).

Problem Statement I

Program Eq. (B2) so that the effects of changing K_c can be observed. Assume that K_c will take on the values 1, 8, and 27; that the initial concentration in the effluent is $y(0) = 0$; and that $r(t)$ is a unit step function (i.e., $r = 1$ for $t \geq 0$).

APPENDIX B. TUBULAR CHEMICAL REACTOR CONTROL SYSTEM

SOLUTION—PART I

Rewrite equation (B2) as

$$-\ddot{y} = (3/T)\ddot{y} + (3/T^2)y + (1/T^3 + aK_cK_T)y - aK_c$$
$$= (3/T)y + (3/T^2)y + (1/t^3)y + aK_c(K_Ty - 1). \quad \text{(B3)}$$

The program is shown in Fig. B3.

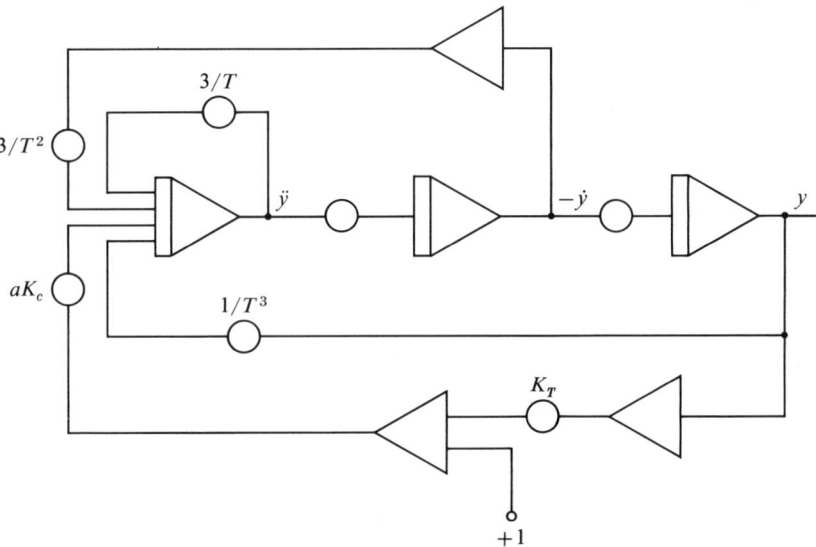

Fig. B3. Unscaled Diagram

PROBLEM STATEMENT II

Now that an unscaled computer diagram has been obtained, scale Eq. (B3) to obtain a completely scaled computer solution.

Since the estimation of maxima has not been considered yet, the maximum values will be given. The determination of these values is presented in the next section.

Assume that

$$|y| \max = 2 \times 10^{-3}$$
$$|\dot{y}| \max = 10 \times 10^{-3}$$
$$|\ddot{y}| \max = 50 \times 10^{-3}.$$

Solution—Part II

Rewriting Eq. (B3)

$$-\dddot{y} = (3/T)\ddot{y} + (3/T^2)\dot{y} + (1/T^3)y + aK_c(K_T y - 1).$$

We see from Fig. B3 that the quantities to be scaled are y, $\dot{y}$, $\ddot{y}$, and $(K_T y - 1)$, as these are the outputs of amplifiers.

Although maximum values for the first three were given, it will be valuable to show how they were estimated here. Recall that the maximum values of the m-th order derivative in an n-th order differential equation is approximated by

$$y_{max}^{(m)} \approx y_{max} \, \omega_n^m$$

where

$$\omega_n = \sqrt[n]{a_o/a_n}.$$

For this problem then,

$$\dot{y}_{max} = \omega_n y_{max} \quad \text{and}$$
$$\ddot{y}_{max} = \omega_n^2 y_{max}$$

since
$$\omega_n = \sqrt[n]{a_o/a_n}; \quad n = 3$$

and
$$a_o = aK_c K_T + 1/T^3 = (1 + K_v K_c K_T K)/T^3$$
$$a_3 = 1$$

Then
$$\omega_n = \sqrt[3]{\frac{1 + K_v K_c K_T K}{T}} = 2\sqrt[3]{1 + K_c}.$$

The table below indicates the various values for ω under the conditions $K_c = 1$, 8, and 27.

K_c	$1 + K_c$	ω_n
1	2	2.52
8	9	4.16
27	28	6.15

Therefore, an ω_n of about 5 would not be too unreasonable.

Now let us consider the maximum value of y. At the steady state condition $\dddot{y} = \ddot{y} = \dot{y} = 0$, from Eq. (B3)

$$0 = y_{ss}(aK_c K_T + 1/T^3) - aK_c$$

or
$$y_{ss} = \frac{aK_c}{aK_c K_T + 1/T^3} = \frac{.001 K_c}{1 + K_c}.$$

APPENDIX B. TUBULAR CHEMICAL REACTOR CONTROL SYSTEM

Then for $K_c = 1, 8, 27$

$$5 \times 10^{-4} \leq \frac{.001 K_c}{1 + K_c} \leq 10 \times 10^{-4}.$$

Therefore

$$5 \times 10^{-4} \leq y_{ss} \leq 10 \times 10^{-4}.$$

With a step input, the response y will probably not be greater than twice the steady state value. Therefore, a value of $|y| \leq 2 \times 10^{-3}$ will be sufficient.

If one chooses $y_{max} = 2 \times 10^{-3}$ and $\omega_n = 5$, then the following table results:

Physical Variable	R.U.M.	Computer Variable
y	2×10^{-3}	$[500y]$
$\dot{y}$	10×10^{-3}	$[100\dot{y}]$
$\ddot{y}$	50×10^{-3}	$[20\ddot{y}]$
$K_T y - 1$	1	$[K_T y - 1]$

The scaled equation is then developed from the table and eq. (B3).

$$-\frac{d}{dt}[20\ddot{y}]\frac{1}{20} = \left(\frac{3}{T}\right)[20\ddot{y}]\frac{1}{20} + \left(\frac{3}{T^2}\right)[100\dot{y}]\left(\frac{1}{100}\right)$$
$$+ \left(\frac{1}{T^3}\right)[500y]\frac{1}{500} + (aK_c)\left[K_T[500y]\left(\frac{1}{500}\right) - 1\right]$$

$$-\frac{d}{dt}[20\ddot{y}] = (3/T)[20\ddot{y}] + (3/5T^2)[100\dot{y}] + (1/25T^3)[500y]$$
$$+ (20aK_c)[(K_c/500)[500y] - 1] \quad (B4)$$

To time scale, we rewrite Eq. (B4) in terms of τ:

$$-\frac{d}{d\tau}[20\ddot{y}] = (3/T\beta)[20\ddot{y}] + (3/5T^2\beta)[100\dot{y}] + (1/25T^3\beta)[500y]$$
$$+ (20aK_c/\beta)[(K_T/500)[500y] - 1]. \quad (B5)$$

This gives the diagram in Fig. B4.

Inspection of this figure shows the following integrator gains:

(1) $3/T\beta = 6/\beta$
(2, 3) $5/\beta$
(4) $1/25T^2\beta = 0.32/\beta$
(6) $20aK_c/\beta = .16/\beta, 1.28/\beta, 4.32/\beta$
(7) $3/5T^2\beta = 2.4/\beta$

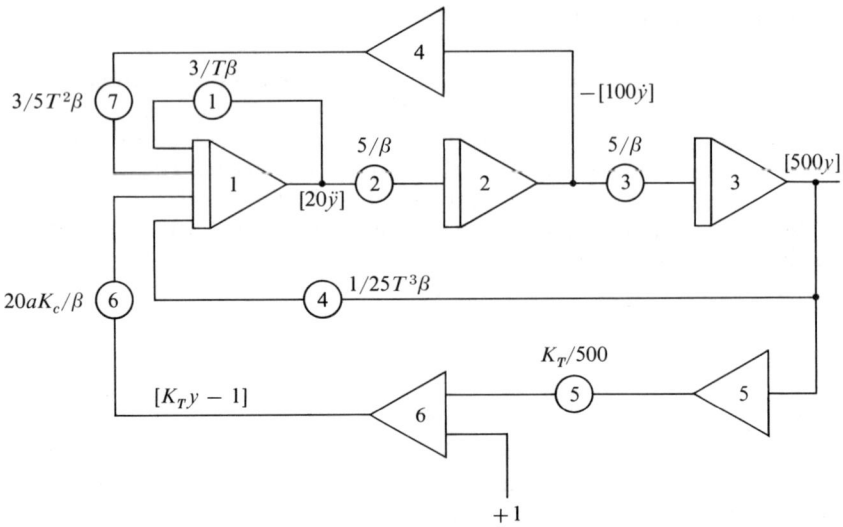

Fig. B4. Amplitude Scaled Program

Thus, $\beta = 1$ appears to be a reasonable value. The final scaled diagram appears in Fig. B5.

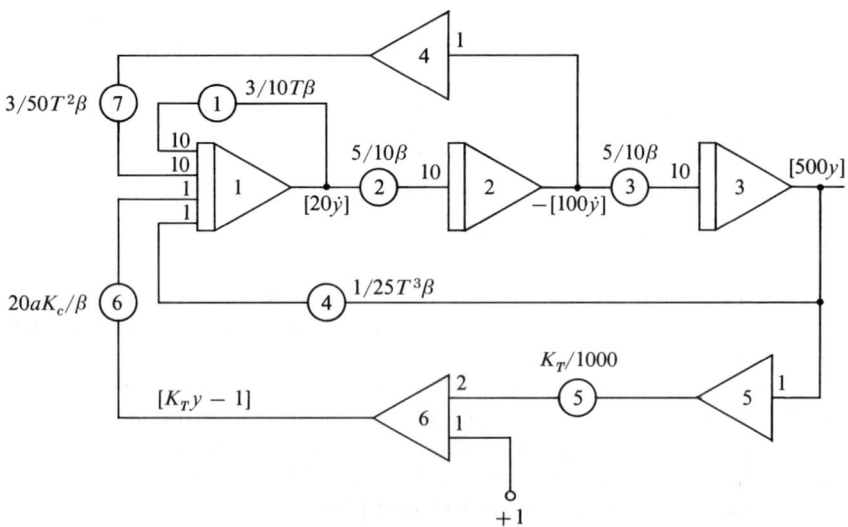

Fig. B5. Scaled Diagram

APPENDIX B. TUBULAR CHEMICAL REACTOR CONTROL SYSTEM 141

PROBLEM STATEMENT III

In the process of programming, scaling and patching a problem, numerous opportunities for error arise. Since even one error in a program can invalidate the entire result, it is absolutely necessary to detect and correct these errors before actual computation starts. The method of detecting and correcting programs is called Static Check.

Perform a static check for the Tubular Chemical Reactor Control System problem where:

$$\ddot{y} = 25 \times 10^{-2}$$
$$\dot{y} = 10 \times 10^{-3}$$
$$y = 2 \times 10^{-3}$$

and $K_c = 2.$

After the static check is completed, patch the problem and observe the effect of changing the value of K_c; that is, $K_c = 1, 8,$ and 27. (Other values of K_c may also be used to determine the effect of changing this parameter.)

SOLUTION—PART III

The performance of a complete static check involves two parts: a *program check* and a *circuit check*. The solution presented here will depict the program check. The circuit check involves patching the program on the computer and measuring voltages to compare to calculations.

For a complete program check, calculations on the computer circuit itself must be made in addition to calculations from the original unscaled equations.

(a) CALCULATIONS BASED ON THE PROGRAM

Calculations based on the program can be performed directly on a copy of the circuit diagram if the outputs of all integrators and all pot settings are known.

The integrator outputs are $[20\ddot{y}]$, $[100\dot{y}]$, and $[500y]$ and their values are found from the values assumed for $\ddot{y}$, $\dot{y}$, and y in the statement of the problem; pot settings are calculated from the assumed values of the parameters (see problem statement). Figure B6 shows the diagram with these quantities marked in appropriate places.

Now, without any other information except that which is shown in Fig. B6, all amplifier outputs can be calculated. Also, all derivative inputs

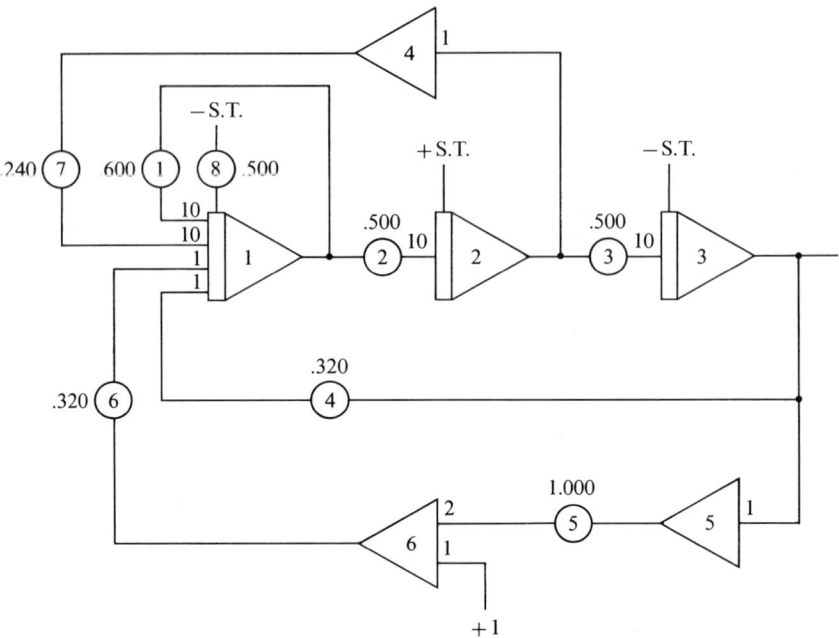

Fig. B6. Static Check

of integrators can be found. For example:

$$A03 = -A05 \qquad\qquad\qquad = 1.000$$
$$A06 = -[2(P05)(A05) + 1] = +1.000$$
$$A02 = -A04 \qquad\qquad\qquad = -1.000$$
$$A01 = \qquad\qquad\qquad\qquad = +.500$$

and

$$D01 = (A06)(P06) + (A03)(P04) + (A04)(P07)10$$
$$\qquad\quad + (A01)(P01)(10) = \quad 6.040$$
$$D02 = (A01)(P02)10 \quad = \quad 2.500$$
$$D03 = (A02)(P03)10 \quad = -5.000.$$

These values are shown in Fig. B7, the completed program check.

(b) CALCULATIONS BASED ON THE ORIGINAL PROBLEM

Calculations based on the original problem are performed for amplifier outputs and derivative values as follows: The output of each amplifier is known in terms of the problem variables and a scale factor. Knowing the assumed values of these allows the calculation of these outputs without

APPENDIX B. TUBULAR CHEMICAL REACTOR CONTROL SYSTEM

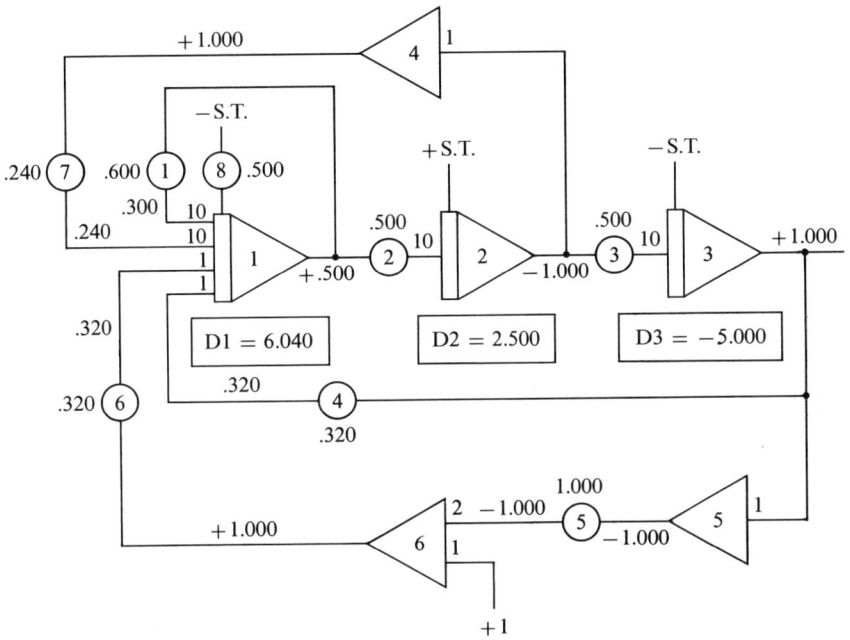

Fig. B7. Static Check

referring to the diagram at all. The scaling table lists the computer variables (i.e., the amplifier outputs) as:

$$A01 = 20\ddot{y} = 20(25 \times 10^{-3}) = 0.500$$
$$A02 = -100\dot{y} = -100(10^{-2}) = 1.000$$
$$A03 = 500y = 500(2 \times 10^{-3}) = +1.000$$
$$A04 = +100\dot{y} = 100(10^{-2}) = +1.000$$
$$A05 = -500y = 500(2 \times 10^{-3}) = -1.000$$
$$A06 = K_T y - 1 = (1000)(2 \times 10^{-3}) - 1 = 1.000.$$

The derivatives may be calculated now. Note that

$$D1 = -\frac{d}{d\tau}[20\ddot{y}] = -20\dddot{y}/\beta$$

$$D2 = -\frac{d}{d\tau}[-100\dot{y}] = +100\ddot{y}/\beta$$

$$D3 = -\frac{d}{d\tau}[500y] = -500\dot{y}/\beta.$$

However, $-\dddot{y}$ can be computed from Eq. (B3) as
$$-\dddot{y} = (3/T)\ddot{y} + (3/T^2)\dot{y} + (aK_cK_T + 1/T^3)y - aK_c.$$
Thus, for $\beta = 1$, $-20\dddot{y}/\beta = +6.04$.
Also, $\ddot{y}$ was selected as 25×10^{-3} so that

$\qquad$ D2 $= 100\ddot{y}/\beta = 2.50$, and $\dot{y}$ was selected as 10^{-2} so that

$\qquad$ D3 $= -500\dot{y}/\beta = -5.00$.

(c) COMPARISONS

Comparisons are now in order. Checking the values calculated in (b) with the ones calculated on the diagram in (a) shows 100% agreement. Thus, it is safe to say that the program of Fig. B3 truly represents the problem described in Part I. These values may now be entered on the pot and amplifier assignment sheets.

(d) FINAL CHECK

The final effort in checking involves patching the program, setting pots, establishing the selected integrator outputs and measuring all amplifier outputs and derivatives. The measured results should agree with the calculated values to better than 1% or an error is indicated.

APPENDIX B. TUBULAR CHEMICAL REACTOR CONTROL SYSTEM 145

POTENTIOMETER ASSIGNMENT SHEET

Date_____   Problem__Tubular Reactor_____

Pot No.	Parameter Description	Setting Static Check	Static Check Output Voltage	Setting Run Number 1	Notes	Pot No.
00						
01	$3/10T\beta$	.600				
02	$5/10\beta$	.500				
03	$5/10\beta$	.500				
04	$1/25T^3\beta$	.320				
05	$K_T/1000$	1.000				
06	$20aK_c/\beta$	.320		.160		
07	$3/50T^2\beta$	.240				
08	$20\ddot{y}(0)$	.500		0		
09						
10						
11						
12						
13						
14						
15						
16						
17						
18						
19						
20						
21						
22						
23						
24						

146 ANALOG/LOGIC COMPUTER PROGRAMMING AND SIMULATION

AMPLIFIER ASSIGNMENT SHEET

Date _____   Problem __Tubular Reactor__

Amp No.	FB	Output Variable	Static Check				Notes
			Calculated		Measured		
			Check Pt.	Output	Check Pt.	Output	
00							$\beta = 1$
01	$\int$	$20\ddot{y}$	$-.604*$	$+.500$			*Check Ampl. has Gain $= -1/10$.
02	$\int$	$-100\dot{y}$	$-.250*$	-1.000			
03	$\int$	$500y$	$+.500*$	$+1.000$			
04	-1	$100\dot{y}$		$+1.000$			
05	-1	$-500y$		-1.000			
06	Σ	$K_T y - 1$		$+1.000$			
07							
08							
09							
10							
11							
12							
13							
14							
15							
16							
17							
18							
19							
20							

APPENDIX C

Analysis of Tapered Nozzle

Introduction

One is often interested in studying the flow of a fluid in a container. This problem considers the simple case of the steady flow of an incompressible fluid through a uniformly tapered nozzle as shown in Fig. C1.

This problem is interesting from a programmer's point of view because it illustrates the computer solution of a differential equation containing a function of the independent variable and also demonstrates the use of relay comparators.

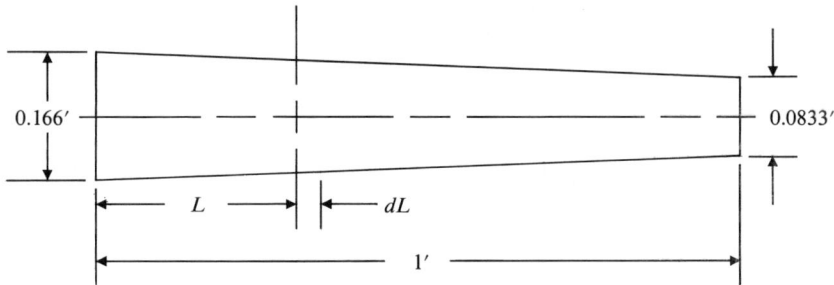

Fig. C1. Tapered Nozzle

The basic energy equation, differential form, for steady, horizontal flow along a streamline is

$$\frac{1}{\gamma}\frac{dp}{dL} + \frac{V}{g}\frac{dV}{dL} = 0 \tag{C1}$$

where

γ = density in pounds per cubic foot
$\gamma = 62.4$ lb./ft.^{3}
p = pressure in pounds per square foot
V = velocity along a streamline in feet per second
g = acceleration due to gravity
$g = 32$ ft./sec.^{2}
$Q = VA$ = discharge rate
$0.4 \leq Q \leq 1.2$ ft.^{3}/sec.

If it is desired to obtain computer-output graphs of the pressure variation along the nozzle for various discharge rates Q, the following equation results from Eq. (C1).

$$\frac{dp}{dL} = -\frac{(1.305) \times 10^5 Q^2}{(2-L)^5} \qquad (C2)$$

This differential equation describing the physical system requires the generation of the function of the independent variable for its solution.

Equation (C2) was obtained from Eq. (C1) where:

$$V = \frac{Q}{A} = \frac{576 Q}{\pi (2-L)^2}$$

$$\frac{dV}{dL} = \frac{1152 Q}{\pi (2-L)^3}; \quad \gamma = 62.4 \text{ lb./ft.}^3$$

PROBLEM STATEMENT I

From Eq. (C2), develop computer circuit, using exactly 7 amplifiers to solve for the pressure (p) as a function of length (L). Recall that for t this simulation $dL = dt$.

PROBLEM SOLUTION I

The direct programming of Eq. (C2) involves the simulation of implicit functions. That is, let $Z = (2 - L)^{-5}$ then differentiating with respect to time,

$$\frac{dZ}{dt} = +5(2-L)^{-6} \qquad (C3)$$

or

$$\frac{dZ}{dt} = \frac{+5(2-L)^{-5}}{2-L} = \frac{+5Z}{2-L}. \qquad (C4)$$

This equation alone cannot be solved implicitly so another equation is needed, that is, let $y = 2 - L$. Implicit differentiation of this equation yields

$$\frac{dy}{dt} = -1 \qquad (C5)$$

APPENDIX C. ANALYSIS OF TAPERED NOZZLE

so that

$$\frac{dZ}{dt} = \frac{+5Z}{2-L} = \frac{+5Z}{y}. \tag{C6}$$

Equation (C6) may be solved using a division circuit and the variable Z may be substituted into Eq. (C3) for the complete solution, that is:

$$dL = dt$$

$$\frac{dp}{dt} = -(1.305) \times 10^5 Q^2 Z. \tag{C7}$$

Figure C2 indicates the unscaled computer diagram for the solution of this problem. Figure C2a is the unscaled computer diagram in symbolic notation and Fig. C2b is the complete circuit with amplifiers indicated.

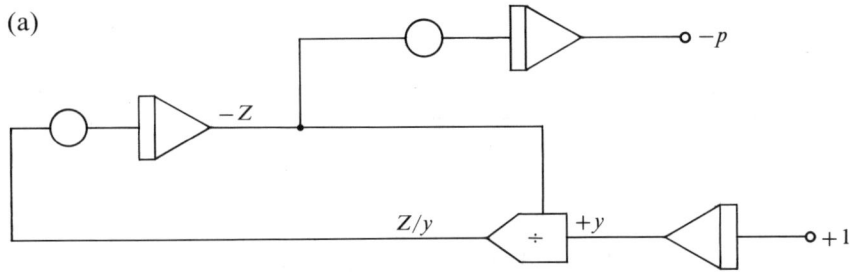

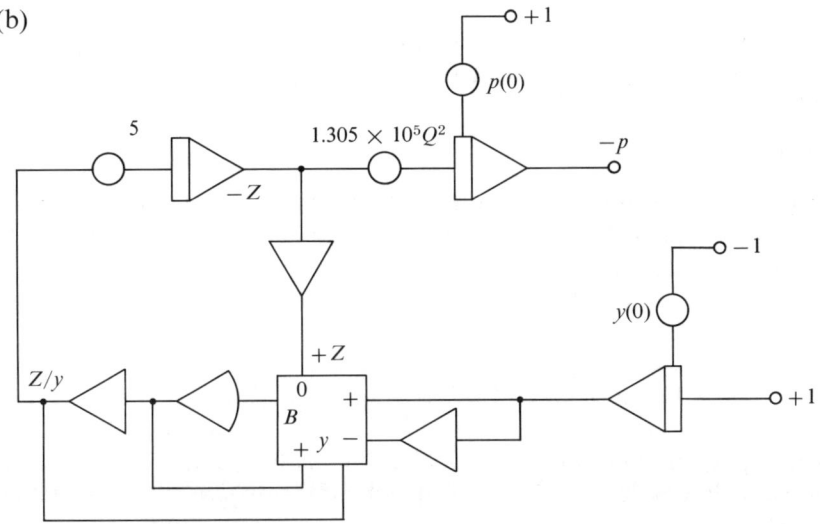

Fig. C2. Unscaled Computer Diagram Showing All Amplifiers (See Computer Reference Handbook for Patching Details)

Problem Statement II

The tapered nozzle problem was programmed above without scaling; now the problem should be amplitude and time scaled.

The maximum values of the variables should be estimated from the data presented previously.

In obtaining a solution, the initial pressure $p(0)$, is determined by a trial-and-error adjustment until a value is found at which the pressure goes to zero at the time corresponding to $L = 1.0$ ft.

It should be pointed out that extreme care should be exercised while setting initial voltages, potentiometer settings, and so on, and in generating the function $Z = (2 - L)^{-5}$. This function, because of its range of values, is characterized by low voltages over a considerable portion of the range.

Solution to Problem Statement II

The maximum values of the outputs of amplifiers can be determined from the equations for y, Z and dp/dL.

That is:
$$y = 2 - L$$
$$Z = (2 - L)^{-5}$$
$$\frac{dp}{dL} = -(1.305) \times 10^5 Q^2 Z.$$

The maximum value of y will occur when $L = 0$, therefore:
$$|y_{max}| = 2 \text{ ft.}$$

The maximum value of Z occurs when $L = 1$, therefore:
$$Z = (2 - 1)^{-5} = 1$$
$$|Z_{max}| = 1 \text{ ft.}$$

The maximum value of p will occur at $L = 0$ and should be zero at $L = 1$. From experimentation it has been found that $|p_{max}| = 50 \times 10^3$ lb./ft.^{2}.

It is now necessary to make up a table with the computer variables including the outputs of all amplifiers as indicated below.

Physical Variable	Rounded up Maximum	Computer Variable
z	2	$[Z/2]$
y	2	$[y/2]$
p	50×10^3	$[p/5 \times 10^4]$

APPENDIX C. ANALYSIS OF TAPERED NOZZLE

The computer variables are now substituted into Eq. (C7) to obtain the scaled equations.

Since $dL = dt$

$$\dot{p} = -(1.305) \times 10^5 Q^2 Z \qquad (C8)$$

$$5 \times 10^4 \frac{d}{dt}\left[\frac{p}{5 \times 10^4}\right] = -(1.305) \times 10^5 Q^2 \left[\frac{Z}{2}\right]^2$$

$$\frac{d}{dt}\left[\frac{p}{5 \times 10^4}\right] = -(1.305) 2 Q^2 \left[\frac{Z}{2}\right]^2$$

$$\frac{d}{dt}\left[\frac{p}{5 \times 10^4}\right] = -(5.220 Q^2)[Z]. \qquad (C9)$$

Equation (C9) is the properly scaled equation; however, one must look at the integrator gains to be certain that they are between 0.1 and 10. To do this, look at the maximum value of the factor multiplying [Z].

$$5.220 Q_{max}^2 = 5.22(1.2)^2 = 7.28$$

Hence, the following scaled equation results:

$$\frac{d}{dt}\left[\frac{p}{5 \times 10^4}\right] = -10\left(\frac{5.22 Q^2}{10}\right)\left[\frac{Z}{2}\right]. \qquad (C10)$$

Since the potentiometer settings are between 0.1 and 1.00 and the gain settings are between 0.1 and 10, no time scaling is necessary, that is, $\beta = 1$.

The completely scaled equation is indicated below:

$$\frac{d}{d\tau}\left[\frac{p}{5 \times 10^4}\right] = -10\left(\frac{5.22 Q^2}{10\beta}\right)\left[\frac{Z}{2}\right]. \qquad (C11)$$

In order to completely scale the problem Eqs. (C5) and (C6) must be scaled also:

$$2\frac{d}{dt}\left[\frac{y}{2}\right] = -1$$

$$\frac{d}{dt}\left[\frac{y}{2}\right] = -1/2$$

$$\frac{d}{d\tau}\left[\frac{y}{2}\right] = -1/2\beta. \qquad (C12)$$

Equation (C13) is the one implemented on the computer.

$$\frac{-d}{d\tau}\left[\frac{y}{2}\right] = 1\left(\frac{1}{2\beta}\right)[1] \qquad (C13)$$

For Eq. (C6):

$$\frac{dZ}{dt} = +5\frac{Z}{y} \qquad (C14)$$

$$2\frac{d}{dt}\left[\frac{Z}{2}\right] = +5\frac{[Z/2]}{[y/2]}, \text{ where } \frac{[Z/2]}{[y/2]} = [X] \tag{C15}$$

$$\frac{d}{dt}\left[\frac{Z}{2}\right] = +2.5[X] \tag{C16}$$

$$\frac{d}{d\tau}\left[\frac{Z}{2}\right] = +\frac{10(0.25)}{\beta}[X]. \tag{C17}$$

If one looks closely at Eq. (C15) it can be seen that for $y/2 < 0.5$, that is, for $y = 2 - L$, $L > 1.0$ the quotient of $[Z/2]/[y/2] = [X] > 1$ will become too large. Since $L > 1.0$ corresponds to going past the end of the nozzle in addition to overloading the circuit, the solution will have to be held at $[y/2] = 0.5$. At the end of the nozzle, where $L = 1$, a comparator switch is used to hold the input to the integrator at 0.5.

Figure C3 indicates the scaled computer diagram with the switching scheme present. The comparator will transfer to the − contact when $[y/2] \geq 0.5$.

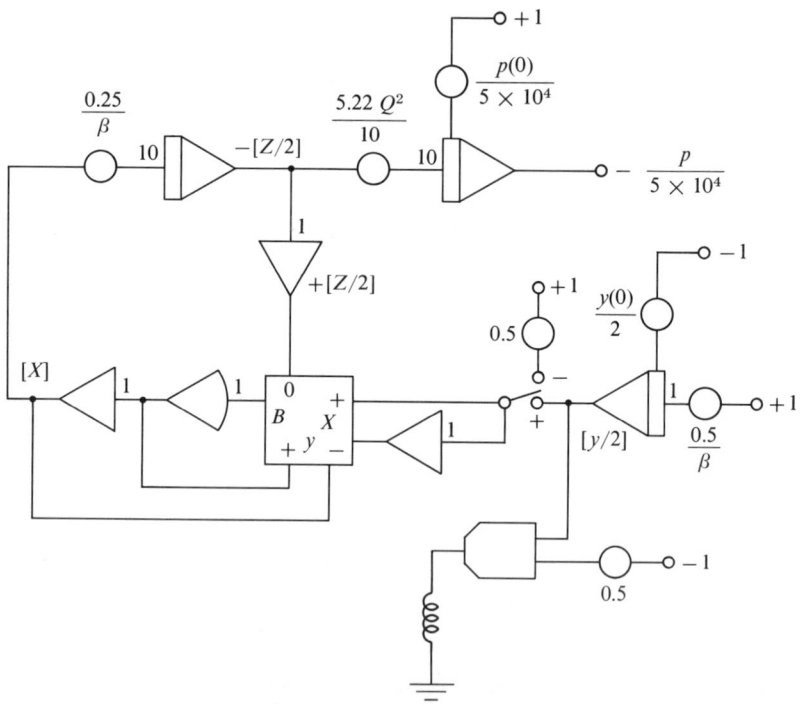

Fig. C3. Scaled Computer Diagram $\beta = 1$

APPENDIX C. ANALYSIS OF TAPERED NOZZLE   153

PROBLEM STATEMENT III

In the process of programming, scaling and patching a problem, numerous opportunities for error arise. Since even one error in a program can invalidate the entire result, it is absolutely necessary to detect and correct these errors before actual computation starts. The method of detecting and correcting programs is called Static Check.

Perform a Static Check for the tapered nozzle where

$$p = 0.5 \times 10^4 \text{ lb./ft.}^2$$
$$Z = (1)^{-5} \text{ ft.}$$
$$y = 2 \text{ ft.}$$
$$Q = 0.6 \text{ ft.}^3/\text{sec.}$$

The maximum value of Z was chosen because Z has small values in most of its range. After the Static Check is completed, patch the problem and investigate the effect of changing the discharge rate Q. Let Q equal the following values and plot a series of curves for pressure p versus length L.

$$Q_1 = 0.4 \text{ ft.}^3/\text{sec.}$$
$$Q_2 = 0.6 \text{ ft.}^3/\text{sec.}$$
$$Q_3 = 0.8 \text{ ft.}^3/\text{sec.}$$
$$Q_4 = 1.0 \text{ ft.}^3/\text{sec.}$$
$$Q_5 = 1.2 \text{ ft.}^3/\text{sec.}$$

SOLUTION TO PROBLEM STATEMENT III

The performance of a complete Static Check involves two parts: a *program check* and a *circuit check*. The solution presented here will depict the program check. The circuit check involves patching the program on the computer and measuring voltages to compare to calculations.

For a complete *program check*, calculations on the computer itself must be made in addition to calculations from the original unscaled equations.

(a) Calculations based on the program can be performed directly on a copy of the circuit diagram if the outputs of all integrators and all pot settings are known.

The integrator outputs are:

$$[Z/2], [p/5 \times 10^4], [y/2]$$

and their values are found from the values assumed for p, Z, and y in the statement of the problem; pot settings are calculated from the assumed values of the parameters, which is Q in this case. Figure C4 shows the diagram with these quantities marked in appropriate places.

Now, without any other information except that which is shown in Fig.

C4, all amplifier outputs can be calculated. Also, all derivative inputs of integrators can be found. For example,

$$A02 = -0.5$$
$$A03 = -0.1$$
$$A04 = -A02 = +0.500$$
$$A06 = +1.00$$
$$A05 = -A06 = -1.00$$
$$A13 = -|A04|/|A05| = -0.500$$
$$A15 = +0.500$$

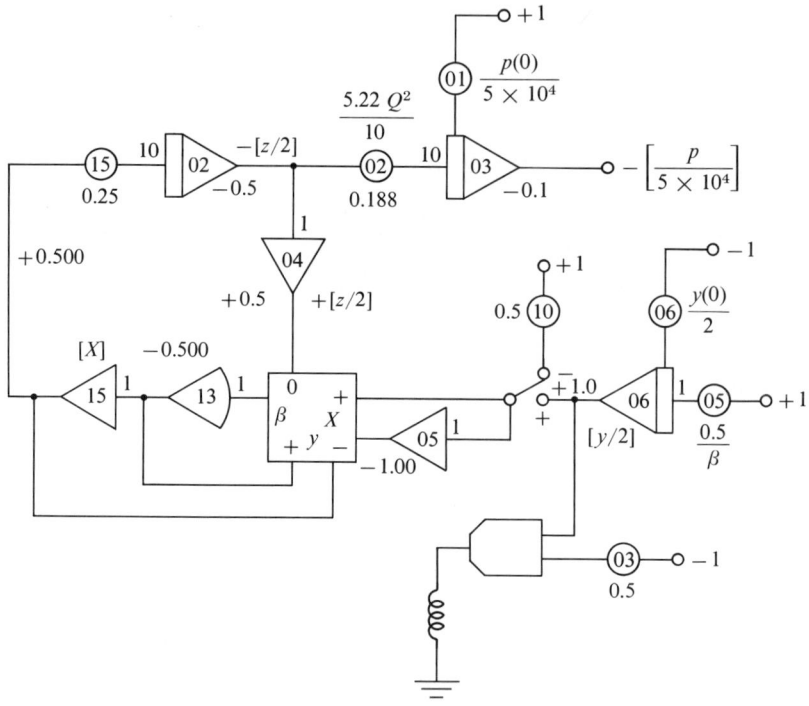

Fig. C4. Static Check

$$D02 = 10[A15](p15)$$
$$D02 = 10[0.5](0.25)$$
$$D02 = +1.25$$
$$D03 = 10(p02)[A02]$$
$$D03 = -10(0.188)[0.5]$$

APPENDIX C. ANALYSIS OF TAPERED NOZZLE

$$D03 = -0.94$$
$$D06 = 1(p05)[+1.0]$$
$$D06 = 1(0.5)[+1.0]$$
$$D06 = +0.500.$$

These values are shown in Fig. C5, the completed program check.

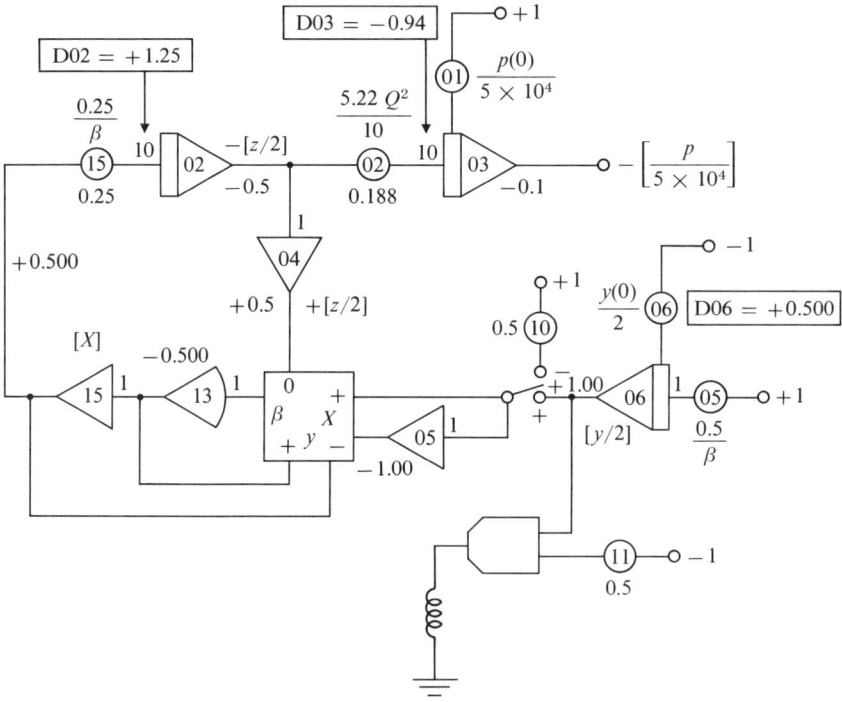

Fig. C5. Calculations for Program Check

(b) Calculations based on the original problem are performed for amplifier outputs and derivative values as follows: The output of each amplifier is known in terms of the problem variables and a scale factor. Knowing the assumed values of these allows the calculation of these outputs without referring to the diagram at all. The scaling table lists the computer variables (i.e., the amplifier outputs) as

$[Z]$
$[p/5 \times 10^4]$
$[y/2]$

and values for Z, p, and y are known from the problem statement. Thus,

$$A02 = -[Z/2] = -[1/2] = -0.500$$

$$A03 = -\left[\frac{p}{5 \times 10^4}\right] = -\left[\frac{0.5 \times 10^4}{5 \times 10^4}\right] = -0.1$$

$$A06 = [y/2] = [2/2] = +1.00$$

$$A04 = +[Z/2] = +0.500$$

$$A05 = -[y/2] = -[2/2] = -1.00$$

$$A13 = -\left[\frac{Z/2}{y/2}\right] = -[0.5/1] = -0.5$$

$$A15 = +\left[\frac{Z/2}{y/2}\right] = +[0.5/1] = +0.500$$

$$D02 = -\frac{d}{d\tau} - [Z/2] = \frac{\dot{Z}/2}{\beta} = +2.5\frac{[Z/2]}{[y/2]}$$

$$D02 = +(2.5)1/2 = +1.25$$

$$D03 = -\frac{d}{d\tau} - \left[\frac{p}{5 \times 10^4}\right] = \frac{1}{\beta}\left[\frac{\dot{p}}{5 \times 10^4}\right] = -(2.61)Q^2[Z]$$

$$D03 = -(5.22)(0.6)^2[1]$$

$$D03 = -0.94$$

$$D06 = -\frac{d}{d\tau}[y/2] = -\beta[\dot{y}/2]$$

$$[\dot{y}/2] = -1/2$$

$$D06 = +0.500.$$

(c) Comparisons are now in order. Checking the values calculated in (b) with the ones calculated on the diagram in (a) shows 100% agreement. Thus, it is safe to say that the program in Fig. C3 truly represents the problem described in Part I. These values may now be entered on pot and amplifier sheets.

(d) The final effort in checking involves patching the program, setting pots, establishing the selected integrator outputs, and measuring all amplifier outputs and derivatives. The measured results should agree with the calculated values to better than 1% or an error is indicated.

Proper integrator outputs for checking this problem are assured by the addition of test IC's for the three integrators through pots 00, 01, and 06.

APPENDIX C. ANALYSIS OF TAPERED NOZZLE

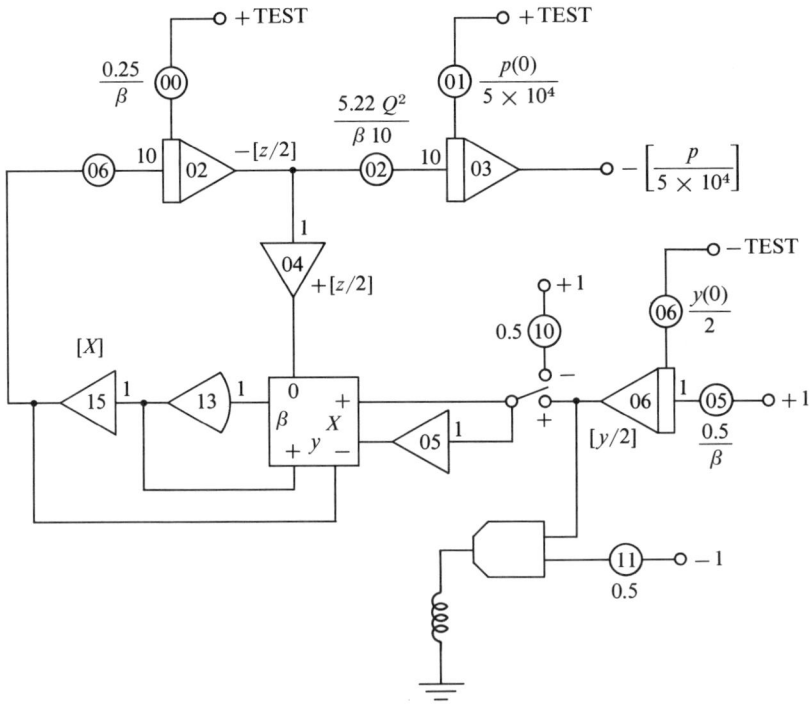

Fig. C6. Final Computer Diagram

POTENTIOMETER ASSIGNMENT SHEET

Date _____   Problem __Tapered Nozzle__

Pot. No.	Parameter Description	Setting Static Check	Static Check Output Voltage	Setting Run Number 1	Notes	Pot. No.
00	$Z(0)$	1.000		0.000		
01	$p(0)/5 \times 10^4$	0.100		—	*	
02	$5.22\ Q^2/10$	0.939			$Q = 0.4, 0.6, 0.8,$	
03					$1.0, 1.2$	
04						
05	0.5	0.5		0.5		
06	$y(0)/2$	1.00				
07						
08					*$p(0)$ is deter-	
09					mined by a trial-and-error	
10	0.5	0.5		0.5	adjustment until a value is found	
11	0.5	0.5		0.5	at which the pressure goes	
12					to zero at the time $L = 1.0$ ft.	
13					that is:	
14				.	(1) Set A06 = $+0.5$	
15	0.25	0.25		0.25	(2) Set A02 = -0.5	
16					(3) Adjust P01 for ≈ 0.0	
17					output.	
18						
19						
20						
21						
22						
23						
24						

APPENDIX C. ANALYSIS OF TAPERED NOZZLE 159

AMPLIFIER ASSIGNMENT SHEET

Date _____   Problem __Tapered Nozzle__

| Amp. No. | FB | Output Variable | Static Check ||||Notes |
| | | | Calculated || Measured || |
			Check Pt.	Output	Check Pt.	Output	
00							
01							
02	$\int$	$-[Z/2]$	$-0.125*$	-1.00			
03	$\int$	$-[p/5 \times 10^4]$	-0.94	-0.100			*Check Amplifier Gain $= -1/10$
04	inv.	$+[z/2]$		$+1.00$			
05	inv.	$-[y/2]$		-1.00			
06	$\int$	$+[y/2]$	-0.500	$+1.00$			
07							
08			Measured through Check Amplifier				
09							
10							
11							
12							
13	H.G.	$-[z/2]/[y/2]$		-1.00			
14							
15	inv.	$+[z/2]/[y/2]$		$+1.00$			
16							
17							
18							
19							
20							

APPENDIX D

Electron Ballistics

CONSIDER THE PROBLEM of obtaining the trajectory of an electron under the influence of combined magnetic and electric fields as shown in Fig. D1.

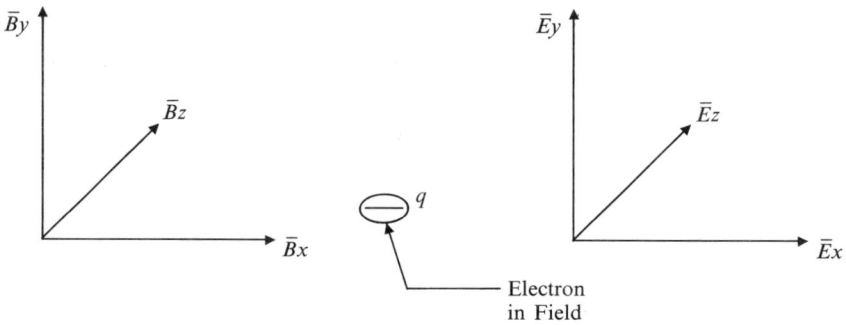

Fig. D1.

The force acting on the electron is given by the vector equation

$$\bar{F} = q(\bar{E} + \bar{V} \times \bar{B})$$

where

q = charge of electron
$\bar{E}$ = electric field vector
$\bar{V}$ = velocity of the particle
$\bar{B}$ = magnetic field vector.

If one wishes to study the effects produced by an orthogonal field relationship, the problem could then be simplified by letting the magnetic

APPENDIX D. ELECTRON BALLISTICS

field have a component only in the Z-direction ($\bar{B}x = \bar{B}y = 0$) and by setting $\bar{E}z = 0$. The equations of motion are then:

$$\ddot{X} = \frac{q}{m}(\bar{E}x + \bar{B}z\dot{Y}) \tag{D1}$$

$$\ddot{Y} = \frac{q}{m}(\bar{E}y - \bar{B}z\dot{X}). \tag{D2}$$

If the magnetic field is constant, the equations are linear ordinary differential equations with constant coefficients.

With this system of equations, the effects of constant, alternating, or rotating fields can be studied. For this study, it shall be assumed that $\bar{E}x$ and $\bar{E}y$ are constant electric fields.

Problem Statement I

Program Eqs. (D1) and (D2) so that the effects of adding or deleting fields as well as changing initial conditions can be investigated. That is, the $\bar{B}$ fields as well as the $\bar{E}$ fields should be isolated and $X(o) = C_1$, $Y(o) = C_2$, $\dot{X}(o) = C_3$, $\dot{Y}(o) = C_4$. Exactly seven (7) amplifiers should be used.

Solution to Problem Statement I

Rewriting Eqs. (D1) and (D2) the following equations are obtained:

$$-\ddot{X} = \frac{q}{m}(-\bar{E}x - \bar{B}z\dot{Y}) \tag{D3}$$

$$-\ddot{Y} = \frac{q}{m}(-\bar{E}y + \bar{B}z\dot{X}). \tag{D4}$$

The unscaled computer diagram is shown in Fig. D2.

Problem Statement II

Now that an unscaled computer diagram has been obtained, scale Eqs. (D3) and (D4) to obtain a completely scaled computer solution.
Assume that:

$$|Bz| \text{ max} = 5{,}000 \text{ Gauss}$$
$$|E| \text{ max} = 5 \times 10^5 \text{ volts/cm}.$$

The mass-charge ratio of an electron is

$$\frac{m}{q} = 5.68 \times 10^{-10} \text{ g/coulomb}$$
$$= 5.68 \times 10^{-16} \text{ volt sec.}^2/\text{cm}^2$$

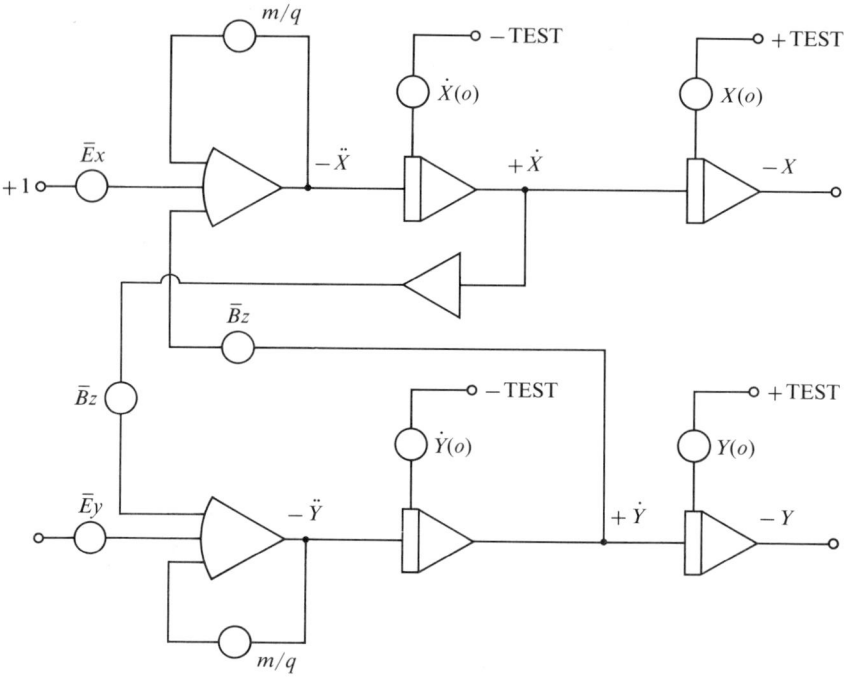

Fig. D2. Unscaled Diagram

where:

$$1 \text{ Gauss} = 10^{-8} \text{ volts sec./cm}^2$$
$$|\dot{X}_{max}| = |\dot{Y}_{max}| = 10^{10} \text{ cm/sec.}$$
$$|\ddot{X}_{max}| = |\ddot{Y}_{max}| = 10^{21} \text{ cm/sec.}^2$$
$$X_{max} = Y_{max} = 1 \text{ cm.}$$

These maxima may have to be changed to meet varying initial conditions and field strengths.

SOLUTION TO PROBLEM STATEMENT II

Since all of the maximum values are given, one need not estimate them. A table indicating the outputs of amplifiers is included below.

Physical Variable	R.U.M.	Computer Variable
X	1	$[X]$
Y	1	$[Y]$
$\dot{X}$	10^{10}	$[\dot{X}/10^{10}]$
$\dot{Y}$	10^{10}	$[\dot{Y}/10^{10}]$

APPENDIX D. ELECTRON BALLISTICS

The scaled equations may then be written as:

$$\frac{d}{dt}\left[\frac{\dot{x}}{10^{10}}\right]10^{10} = \frac{q}{m}\left(\bar{E}x + \bar{B}z\left[\frac{\dot{y}}{10^{10}}\right]10^{10}\right) \quad \text{(D5)}$$

$$\frac{d}{dt}\left[\frac{\dot{y}}{10^{10}}\right]10^{10} = \frac{q}{m}\left(\bar{E}y - \bar{B}z\left[\frac{\dot{x}}{10^{10}}\right]10^{10}\right) \quad \text{(D6)}$$

$$\frac{d}{dt}\left[\frac{\dot{y}}{10^{10}}\right] = \left(\frac{q}{m}\right)\left(\frac{\bar{E}x}{10^{10}} + \bar{B}z\left[\frac{\dot{y}}{10^{10}}\right]\right) \quad \text{(D7)}$$

$$\frac{d}{dt}\left[\frac{\dot{y}}{10^{10}}\right] = \left(\frac{q}{m}\right)\left(\frac{\bar{E}y}{10^{10}} - \bar{B}z\left[\frac{\dot{x}}{10^{10}}\right]\right). \quad \text{(D8)}$$

One needs to examine the integrator gains to be assured they are between 0.1 and 1. If they are not, time scaling is necessary.

Since the $\bar{B}$ and $\bar{E}$ fields are isolated on potentiometers, Eqs. (D7) and (D8) may be written with the maximum values of these potentiometers indicated. The time scaled equations will also be included: (The maximum value of βz has been converted to volts sec/cm).

$$\frac{d}{d\tau}\left[\frac{\dot{x}}{10^{10}}\right] = \frac{1}{\beta\left(\frac{m}{q}\right)}\left\{\left(\frac{\bar{E}x}{5 \times 10^5}\right)\frac{0.5 \times 10^6}{10^{10}}\right.$$
$$\left. + \left(\frac{\bar{B}z}{5 \times 10^3}\right)0.5 \times 10^{-4} \times \left[\frac{\dot{y}}{10^{10}}\right]\right\} \quad \text{(D9)}$$

$$\frac{d}{d\tau}\left[\frac{\dot{y}}{10^{10}}\right] = \frac{1}{\beta\left(\frac{m}{q}\right)}\left\{\left(\frac{\bar{E}y}{5 \times 10^5}\right)\frac{0.5 \times 10^6}{10^{10}}\right.$$
$$\left. - \left(\frac{\bar{B}z}{5 \times 10^3}\right)0.5 \times 10^{-4} \times \left[\frac{\dot{x}}{10^{10}}\right]\right\} \quad \text{(D10)}$$

Since $\frac{m}{q} = 0.568 \times 10^{-15}$

$$\frac{d}{d\tau}\left[\frac{\dot{x}}{10^{10}}\right] = \frac{10}{2(10)\beta(0.568 \times 10^{-15})}\left\{\left(\frac{\bar{E}x}{5 \times 10^5}\right)\frac{1}{10^4}\right.$$
$$\left. + \left(\frac{\bar{B}z}{5 \times 10^3}\right) \times \frac{1}{10^4}\left[\frac{\dot{y}}{10^{10}}\right]\right\} \quad \text{(D11)}$$

$$\frac{d}{d\tau}\left[\frac{\dot{y}}{10^{10}}\right] = \frac{10}{2(10)\beta(0.568 \times 10^{-15})}\left\{\left(\frac{\bar{E}y}{5 \times 10^5}\right)\frac{1}{10^4}\right.$$
$$\left. - \left(\frac{\bar{B}z}{5 \times 10^3}\right) \times \frac{1}{10^4}\left[\frac{\dot{x}}{10^{10}}\right]\right\}. \quad \text{(D12)}$$

The resulting time scaled equations for $\beta = 10^{10}$ are:

$$\frac{d}{d\tau}\left[\frac{\dot{x}}{10^{10}}\right] = \frac{10}{2(0.568)}\left\{\left(\frac{\bar{E}x}{5 \times 10^5}\right) + \left(\frac{\bar{B}z}{5 \times 10^3}\right)\left[\frac{\dot{y}}{10^{10}}\right]\right\} \quad \text{(D13)}$$

$$\frac{d}{d\tau}\left[\frac{\dot{y}}{10^{10}}\right] = \frac{10}{2(0.568)}\left\{\left(\frac{\bar{E}y}{5 \times 10^5}\right) - \left(\frac{\bar{B}z}{5 \times 10^3}\right)\left[\frac{\dot{x}}{10^{10}}\right]\right\}. \quad (D14)$$

For $\beta = 10^{10}$, 1 second of computer time (τ) equals 100 picoseconds of problem time.

The completely scaled Electron Ballistic problem is indicated in Fig. D3.

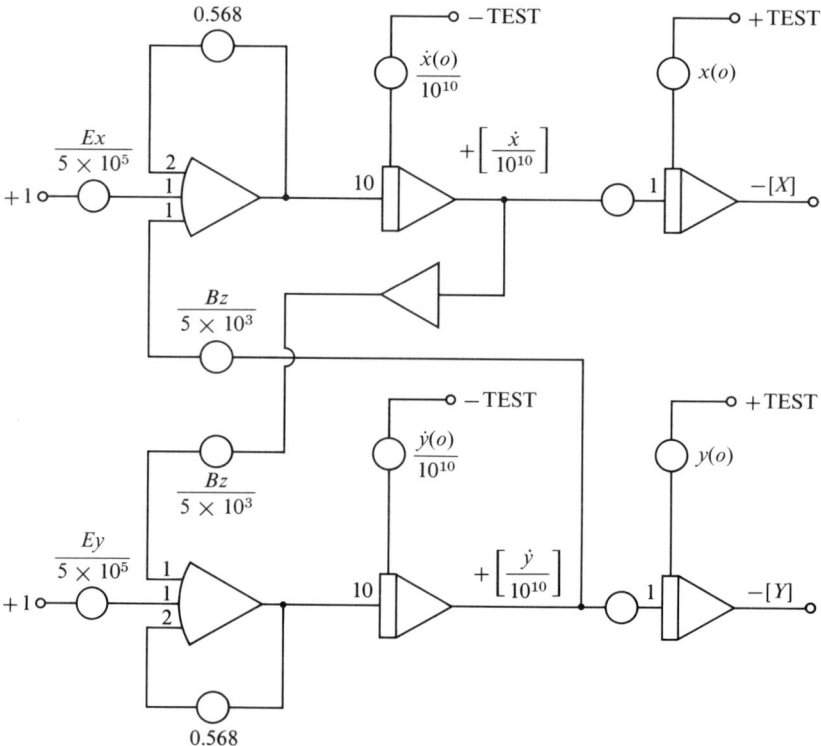

Fig. D3. Completely Scaled Computer Solution ($\beta = 10^{10}$)

PROBLEM STATEMENT III

In the process of programming, scaling and patching a problem, numerous opportunities for errors arise. Since even one error in a program can invalidate the entire result, it is absolutely necessary to detect and correct these errors before actual computation starts. The method of detecting and correcting programs is called Static Check.

APPENDIX D. ELECTRON BALLISTICS

Perform a Static Check for the Automobile Suspension problem where

$$\frac{m}{q} = 5.68 \times 10^{-16} \text{ volt sec./cm}^2$$
$$\bar{B}z = 4 \times 10^{+3} \text{ Gauss}$$
$$\bar{E}x = \bar{E}y = 2.5 \times 10^5 \text{ volts/cm}$$
$$\dot{x} = 10^{10} \text{ cm/sec.}$$
$$\dot{y} = 0.5 \times 10^{10} \text{ cm/sec.}$$
$$x = 0.25 \text{ cm}$$
$$y = 1 \text{ cm.}$$

After the Static Check is completed, patch the problem and investigate the path of the electron for the following conditions:
(1) $\bar{E}x = \bar{E}y = 0$

$$|\dot{x}_o| = 0 \quad |\dot{y}_o| = 0.5$$
$$\bar{B}_{z1} = 1000 \text{ Gauss}$$
$$\bar{B}_{z2} = 2500 \text{ Gauss}$$
$$\bar{B}_{z3} = 5000 \text{ Gauss}$$

(2) Effect of adding Electric Field $\bar{E}y$

$$\bar{E}x = 0 \quad 0 \leq \bar{E}y \leq 1000 \text{ volts/cm}$$
$$\bar{B}_{z1} = 1000 \text{ Gauss}$$
$$\bar{B}_{z2} = 2500 \text{ Gauss}$$
$$\bar{B}_{z3} = 5000 \text{ Gauss}$$

(a) Note the effect of varying $\bar{E}y$ from 0 to 1000 volts/cm.
(b) With a constant $\bar{E}y$ and $\bar{B}z$ note the effect of an initial condition $x(o)$. The inquisitive student may attempt other variations.

SOLUTION TO PROBLEM STATEMENT III

The performance of a complete Static Check involves two parts: a *program check* and a *circuit check*. The solution presented here will depict the program check. The circuit check involves patching the program on the computer and measuring voltages to compare to calculations.

For a complete *program check,* calculations on the computer circuit itself must be made in addition to calculations from the original unscaled equations.

(a) Calculations based on the program can be performed directly on a copy of the circuit diagram if the outputs of all integrators and all pot settings are known.

The integrator outputs are

$$[\dot{x}/10^{10}], \; -[X], \; +[\dot{y}/10^{10}], \; -[Y]$$

and their values are found from the values assumed for $\dot{x}$, x, $\dot{y}$, and y in the statement of the problem; pot settings are calculated from the assumed values of the parameters (see problem statement). Figure D4 shows the diagram with these quantities marked in appropriate places.

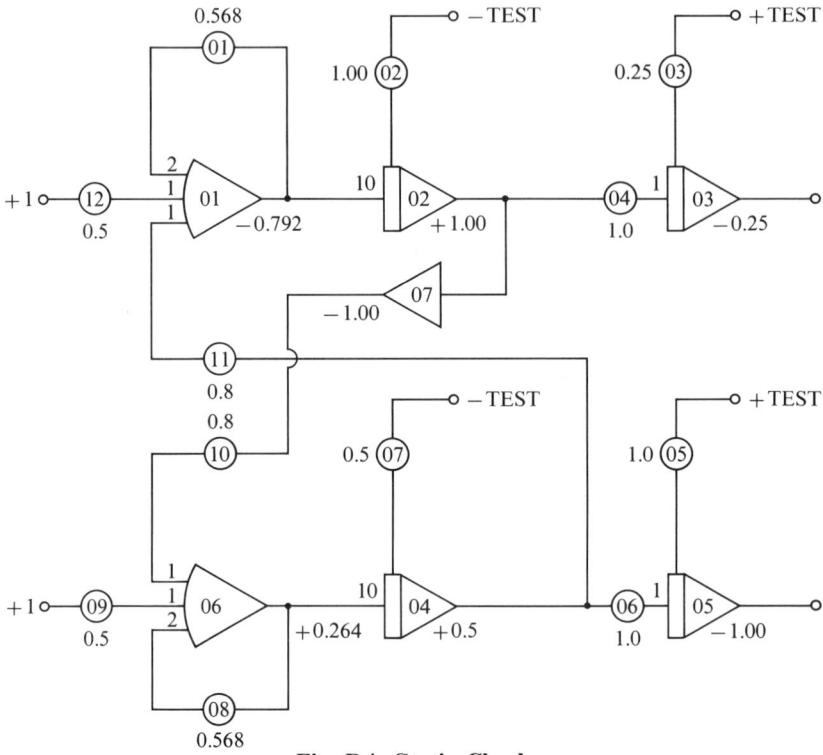

Fig. D4. Static Check

Now, without any other information except that which is shown in Fig. D4, all amplifier outputs can be calculated. Also, all derivative inputs of integrators can be found. For example,

$$A02 = +1.00$$
$$A03 = -0.25$$
$$A04 = +0.5$$
$$A05 = -1.00$$
$$A07 = -1.00$$

$$A01 = \frac{1}{2(P01)}\{-1(P11)[A04] - 1(P12)[+1]\}$$

$$A01 = \frac{1}{2(0.568)}\{-1(0.8)[+0.5] - 1(0.5)[+1]\}$$

$$A01 = -0.792$$

APPENDIX D. ELECTRON BALLISTICS

$$A06 = \frac{1}{2(P08)}\{-1(P10)[A07] - 1(P09)[+1]\}$$

$$A06 = \frac{1}{2(0.568)}\{-1(0.8)[-1] - 1(0.5)[+1]\}$$

$$A06 = +0.264$$

and

$$D02 = 10[A01]$$
$$D02 = 10[-0.792] = -7.92$$
$$D04 = 10[A06] = 10[+0.264] = +2.64$$
$$D03 = 1(P04)[A02]$$
$$D03 = 1(1.0)[+1.0] = +1.00$$
$$D05 = 1(P06)[A04]$$
$$D05 = 1(1.0)[+0.5] = +0.5.$$

These values are shown in Fig. D5, the completed program check.

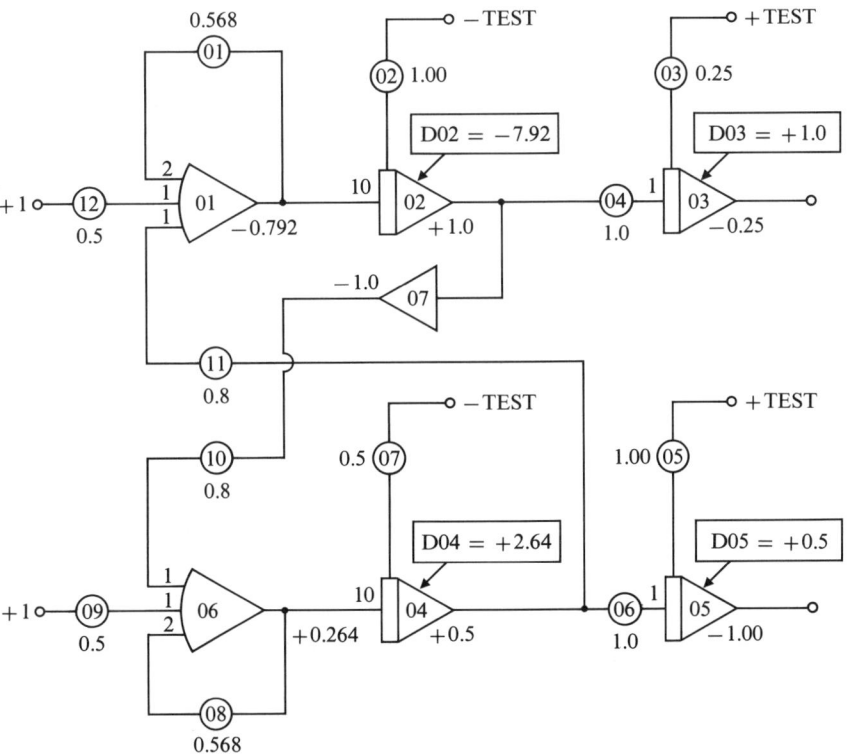

Fig. D5. Complete Static Check

(b) Calculations based on the original problem are performed for amplifier outputs and derivative values as follows: The output of each amplifier is known in terms of the problem variables and a scale factor. Knowing the assumed values of these allows the calculation of these outputs without referring to the diagram at all. The scaling table lists the computer variables (i.e., the amplifier outputs) as

$$[\dot{x}/10^{10}]$$
$$[X]$$
$$[\dot{y}/10^{10}]$$
$$[Y]$$

and values for $x_1, x_2, x_3, \dot{x}_1, \dot{x}_2$ are known from the problem statement. Thus,

$$A02 = [\dot{x}/10^{10}] = [10^{10}/10^{10}] = 1.00$$

$$A03 = -[X] = -[0.25] = -0.25$$

$$A04 = [\dot{y}/10^{10}] = [0.5 \times 10^{10}/10^{10}] = 0.5$$

$$A05 = -[Y] = -[1] = -1.00$$

$$A07 = -A02 = -1.00$$

$$A01 = -\frac{1}{2(0.568)}\{1(0.5)[1] + 1(0.8)[+0.5]\}$$

$$A01 = -0.792$$

$$A06 = -\frac{1}{2(0.568)}\{1(0.8)[-1.0] + 1(0.5)[1]\}$$

$$A06 = +0.264$$

$$D02 = -\frac{d}{d\tau}\left[\frac{\dot{x}}{10^{10}}\right] = \frac{-[\dot{x}/10^{10}]}{\beta} - \frac{d}{d\tau}\left[\frac{\dot{x}}{10^{10}}\right]$$

$$= \frac{-10}{2(0.568)}\left\{\left(\frac{\bar{E}x}{5 \times 10^5}\right) + \left(\frac{\bar{B}z}{5 \times 10^3}\right)\left[\frac{\dot{y}}{10^{10}}\right]\right\}$$

$$= \frac{-10}{1.136}\left\{\left(\frac{2.5 \times 10^5}{5 \times 10^5}\right) + \left(\frac{4 \times 10^3}{5 \times 10^3}\right)\left[\frac{0.5 \times 10^{10}}{10^{10}}\right]\right\}$$

$$= \frac{-10}{1.136}\{(0.5) + (0.8)[0.5]\}$$

$$D02 = -7.92$$

$$D04 = -\frac{d}{d\tau}\left[\frac{\dot{y}}{10^{10}}\right]$$

$$= \frac{-10}{2(0.568)}\left\{\left(\frac{\bar{E}y}{5 \times 10^5}\right) - \left(\frac{\bar{B}z}{5 \times 10^3}\right)\left[\frac{\dot{x}}{10^{10}}\right]\right\}$$

$$D04 = \frac{-10}{1.136}\left\{\left(\frac{2.5 \times 10^5}{5 \times 10^5}\right) - \left(\frac{4 \times 10^3}{5 \times 10^3}\right)\left[\frac{10^{10}}{10^{10}}\right]\right\}$$

$$D04 = \frac{-10}{1.136}\{0.5 - 0.8\}$$

$$D04 = +2.64$$

$$D03 = -\frac{d}{d\tau}[-X] = +\dot{x}/\beta$$

$$D03 = \frac{-10^{10}}{10^{10}} = +1.00$$

$$D05 = -\frac{d}{d\tau}[-Y] = +\dot{y}/\beta$$

$$D05 = \frac{0.5 \times 10^{10}}{10^{10}} = +0.5.$$

(c) Comparisons are now in order. Checking the values calculated in (b) with the ones calculated on the diagram in (a) shows 100% agreement. Thus, it is safe to say that the program of Fig. D3 truly represents the problem described in Part I. These values may now be entered on pot and amplifier sheets.

(d) The final effort in checking involves patching the program, setting pots, establishing the selected integrator outputs, and measuring all amplifier outputs and derivatives. The measured results should agree with calculated values to better than 1% or an error is indicated.

Proper integrator outputs for checking this problem are assured by the addition of test IC's for the four integrators through pots P02, P03, P05 and P07 as shown in Fig. D6.

170 ANALOG/LOGIC COMPUTER PROGRAMMING AND SIMULATION

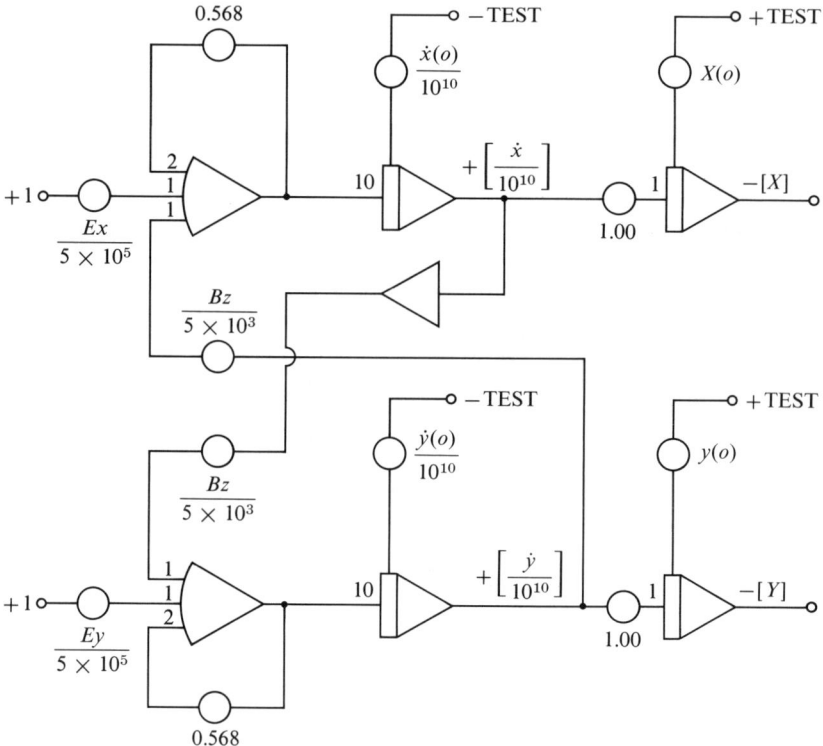

Fig. D6. Final Computer Program

APPENDIX D. ELECTRON BALLISTICS 171

POTENTIOMETER ASSIGNMENT SHEET

Date _____   Problem __Electron Dynamics__

Pot. No.	Parameter Description	Setting Static Check	Static Check Output Voltage	Setting Run Number 1	Notes	Pot. No.
00					$\beta = 10^{10}$	
01	0.568	0.568		0.568		
02	$\dot{x}(0)/10^{10}$	1.00		0.00		
03	$x(o)$	0.25		0.00		
04	$10^{10}/\beta$	1.00		1.00		
05	$y(o)$	1.00		0.00		
06	$10^{10}/\beta$	1.00		1.00		
07	$\dot{y}(o)/10^{10}$	0.500		0.500		
08	0.568	0.568		0.568		
09	$\bar{E}y/5 \times 10^5$	0.500		0.00		
10	$\bar{B}z/5 \times 10^3$	0.800		0.20	Bz varies according $\bar{B}_{z1} = 1000$ Gauss $\bar{B}_{z2} = 2500$ Gauss $\bar{B}_{z3} = 5000$ Gauss	
11	$\bar{B}z/5 \times 10^3$	0.800		0.20		
12	$\bar{E}x/5 \times 10^5$	0.500		0.00		
13						
14						
15						
16						
17						
18						
19						
20						
21						
22						
23						
24						

AMPLIFIER ASSIGNMENT SHEET

Date _____   Problem __Electron Dynamics__

| Amp. No. | FB | Output Variable | Static Check |||| Notes |
| | | | Calculated || Measured || |
			Check Pt.	Output	Check Pt.	Output	
00							
01	H.G.	$-\ddot{x}/10^{10}$		-0.792			
02	$\int$	$[\dot{x}/10^{10}]$	$+0.792^*$	$+1.00$			
03	$\int$	$-[X]$	-0.10^*	-0.25			*Check Amplifier Gain = $-1/10$
04	$\int$	$[\dot{y}/10^{10}]$	-0.264^*	$+0.5$			
05	$\int$	$-[Y]$	-0.50	-1.00			
06	H.G.	$-\ddot{y}/10^{10}$		$+0.264$			
07	inv.	$-\dot{x}/10^{10}$		-1.00			
08			Output of Check Amp.				
09							
10							
11							
12							
13							
14							
15							
16							
17							
18							
19							
20							

APPENDIX E

D.C. Servo Simulation

The d.c. Servo Motor is used extensively in small and medium-sized control systems applications. Two types of operation are commonly used depending on the particular requirements of the load being controlled. Whereas a constant field voltage and varying armature voltage is found suitable for low-speed high-torque operation, a constant armature voltage with v varying field voltage is better for high-speed low-torque operation. The differential equations representing the motor will not be the same for the two different configurations.

This problem depicts the programming considerations for the d.c. Servo Motor. Consider the motor with constant armature voltage and controlled by a variable field voltage as shown in Fig. E1.

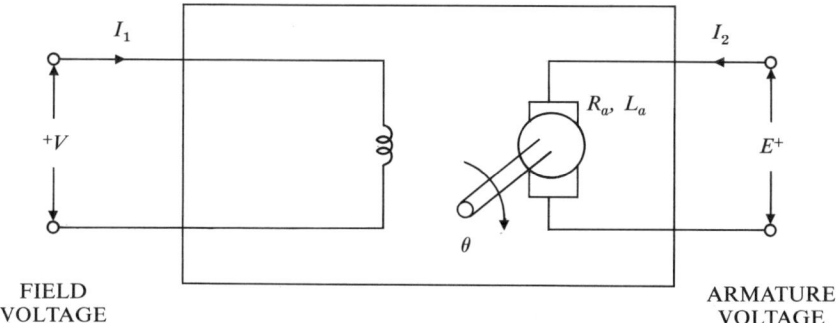

Fig. E1.

The system is described by the following equations:

$$V = R_f I_1 + L_f \dot{I}_1 \tag{E1}$$
$$E = R_a I_2 + L_a \dot{I}_2 + K_v I_1 \dot{\theta} \tag{E2}$$
$$J\ddot{\theta} + F\dot{\theta} = K_T I_1 I_2 \tag{E3}$$

where

V = field voltage (200 V, max.)
R_f = field resistance (50 ohms)
L_f = field inductance (50 henries)
E = armature voltage (500 V, max.)
R_a = armature resistance (4.0 ohms)
L_a = armature inductance (9.0 henries)
I_1 = field airvent (5 amps., max.)
I_2 = armature airvent (80 amps., max.)
K_v = "back-emf" constant (.060 volt-sec./amp.)
K_T = torque constant (.500 ft.-lb./(amp.)^2)
J = motor inertion (5.0 ft.-lb./(rad/sec.^{2}))
F = motor bearing friction (1.0 ft.-lb./(rad/sec.))
θ = armature shaft angle (200 radians, max.).

Problem Statement I

Draw an unscaled computer diagram that represents the d.c. Servo Motor. Exactly eight (8) total amplifiers (and 2 multiplier networks) should be used.

Solution to Problem Statement I

If one solves Eqs. (E1), (E2) and (E3) for the highest order derivative, the following equations result:

$$\frac{dI^1}{dt} = \frac{V}{L_f} - \frac{R_f}{L_f} I_1 \tag{E4}$$

$$\frac{dI_2}{dt} = \frac{E}{L_a} - \frac{R_a}{L_a} I_2 - \frac{K_v}{L_a} I_1 \theta \tag{E5}$$

and

$$\frac{d\theta}{dt} = \frac{K_T I_1 I_2}{J} - \frac{F}{J} \theta. \tag{E6}$$

If one lets

$$T_1 = \frac{L_f}{R_f} \tag{E7}$$

$$T_2 = \frac{L_a}{R_a} \tag{E8}$$

$$T_3 = \frac{J}{f} \tag{E9}$$

APPENDIX E. D.C. SERVO SIMULATION

then the following equations result:

$$-\frac{dI_1}{dt} = -\frac{V}{L_f} + \frac{1}{T_1}I_1 \tag{E10}$$

$$-\frac{dI_2}{dt} = -\frac{E}{L_a} + \frac{1}{T_2}I_2 + \frac{K_v}{L_a}I_1\theta \tag{E11}$$

$$-\frac{d}{dt} = -\frac{K_T I_1 I_2}{J} + \frac{1}{T_3}\theta. \tag{E12}$$

The unscaled computer diagram for these equations is shown in Fig. E2.

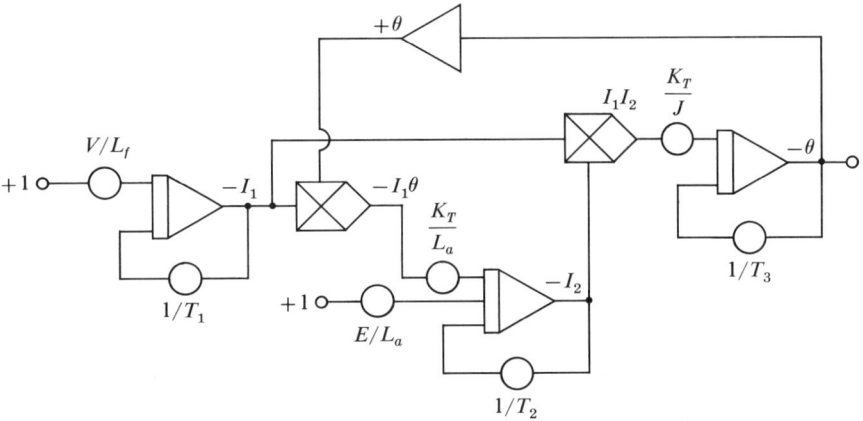

Fig. E2. Unscaled Computer Solution

PROBLEM STATEMENT II

Write the scaled equations and draw a completely scaled computer diagram for the d.c. Servo Motor.

Note the values of the constants and maximum values of the variables in Problem Statement I.

SOLUTION TO PROBLEM STATEMENT II

In order to scale the problem completely one must first construct a computer variable table, such as the one on the following page, indicating the maximum values of the outputs of all amplifiers.

The scaled equations are as follows:

$$\frac{d}{dt}\left[\frac{I_1}{5}\right]5 = \frac{V}{L_f} - \frac{1}{T_1}\left[\frac{I_1}{5}\right]5 \tag{E13}$$

$$\frac{d}{dt}\left[\frac{I_2}{100}\right]100 = \frac{E}{L_a} - \frac{1}{T_2}\left[\frac{I_2}{100}\right]100 - \frac{K_v}{L_a}\left[\frac{I_1\theta}{(5)(200)}\right](200)5 \quad \text{(E14)}$$

$$\frac{d}{dt}\left[\frac{\theta}{200}\right]200 = \frac{K_T}{J}\left[\frac{I_1 I_2}{500}\right]500 + \frac{1}{T_3}\left[\frac{\theta}{200}\right]200 \quad \text{(E15)}$$

which can be written as

$$\frac{d}{dt}\left[\frac{I_1}{5}\right] + \left(\frac{V}{5L_f}\right) - \left(\frac{1}{T_1}\right)\left[\frac{I_1}{5}\right] \quad \text{(E16)}$$

$$\frac{d}{dt}\left[\frac{I_2}{100}\right] = \left(\frac{E}{100 L_a}\right) - \left(\frac{1}{T_2}\right)\left[\frac{I_2}{100}\right] - \left(\frac{10 K_v}{L_a}\right)\left[\frac{I_1\theta}{1000}\right] \quad \text{(E17)}$$

$$\frac{d}{dt}\left[\frac{\theta}{200}\right] = \left(\frac{2.5 K_T}{J}\right)\left[\frac{I_1 I_2}{500}\right] + \left(\frac{1}{T_3}\right)\left[\frac{\theta}{200}\right]. \quad \text{(E18)}$$

Problem Variable	R.U.M.	Computer Variable
I_1	5	$\dfrac{I_1}{5}$
I_2	100	$\dfrac{I_2}{100}$

Examining the maximum value of the coefficients in parentheses;

$$\frac{V}{5L_f} = \frac{200}{5(5)} = 8$$

$$\frac{1}{T_1} = \frac{R_f}{L_f} = \frac{50}{5} = 10$$

$$\frac{E}{100 L_a} = \frac{500}{100(9)} = 0.555$$

$$\frac{1}{T_2} = \frac{L_a}{R_a} = \frac{9}{4} = 2.5$$

$$\frac{10 K_v}{L_a} = \frac{10(0.06)}{9} = 0.0667$$

$$\frac{2.5 K_T}{J} = \frac{2.5(0.5)}{5} = 0.25$$

$$\frac{1}{T_3} = \frac{F}{J} = \frac{1}{5} = 0.2.$$

These integrator gains lie in the range 0.0667 to 10.0. Hence, a reasonable choice of β is $\beta = 1$.

APPENDIX E. D.C. SERVO SIMULATION

Now, time scaling the problem:

$$\frac{d}{d\tau}\left[\frac{I_1}{5}\right] = \left(\frac{V}{5\beta L_f}\right) - \left(\frac{1}{T_1\beta}\right)\left[\frac{I_1}{5}\right] \tag{E19}$$

$$\frac{d}{d\tau}\left[\frac{I_2}{100}\right] = \left(\frac{E}{100\beta L_a}\right) - \left(\frac{1}{T_2\beta}\right)\left[\frac{I_2}{100}\right]$$
$$- \left(\frac{10K_v}{\beta L_a}\right)\left[\frac{I_1\theta}{1000}\right] \tag{E20}$$

$$\frac{d}{d\tau}\left[\frac{\theta}{200}\right] = \left(\frac{2.5K_T}{\beta J}\right)\left[\frac{I_1 I_2}{500}\right] - \left(\frac{1}{T_3\beta}\right)\left[\frac{\theta}{200}\right] \tag{E21}$$

$$\frac{d}{d\tau}\left[\frac{I_1}{5}\right] = 10\left(\frac{V}{50\beta L_f}\right) - 10\left(\frac{1}{10T_1\beta}\right)\left[\frac{I_1}{5}\right] \tag{E22}$$

$$\frac{d}{d\tau}\left[\frac{I_2}{100}\right] = \left(\frac{E}{100\beta L_a}\right) - 10\left(\frac{1}{10T_1\beta}\right)\left[\frac{I_2}{100}\right]$$
$$- 1\left(\frac{10K_v}{\beta L_a}\right)\left[\frac{I_1\theta}{1000}\right] \tag{E23}$$

$$\frac{d}{d\tau}\left[\frac{\theta}{200}\right] = \left(\frac{2.5K_T}{J\beta}\right)\left[\frac{I_1 I_2}{500}\right] - \left(\frac{1}{T_3\beta}\right)\left[\frac{\theta}{200}\right]. \tag{E24}$$

Problem Statement III

In the process of programming, scaling and patching a problem, numerous opportunities for error arise. Since even one error in a program can invalidate the entire result, it is absolutely necessary to detect and correct these errors before actual computation starts. The method of detecting and correcting programs is called Static Check.

Perform a Static Check for the d.c. Servo Motor problem where

$$I_1 = 5 \text{ amps}$$
$$I_2 = 50 \text{ amps}$$
$$\theta = 100 \, \frac{\text{radians}}{\text{sec.}}$$

and parameter values are as in Part I with $v = 100^V$ and $E = 200^V$.

After the Static Check is completed, patch the problem and investigate the effect of (1) introducing a step input, through switch SW I_1 after the problem has been running (set $v = 0$). Try several steps.

(2) Plot the output θ of the d.c. servo system versus time for five different input voltages (V). Be certain to plot θ for $V = 0$ to use as a comparison.

(3) For $J = 1/50$ ft.-lb./rad/sec.^{2} reset the potentiometers to get this fast moving system to slow down. When problem has been rescaled for $B \neq 1$, repeat parts (1) and (2) above.

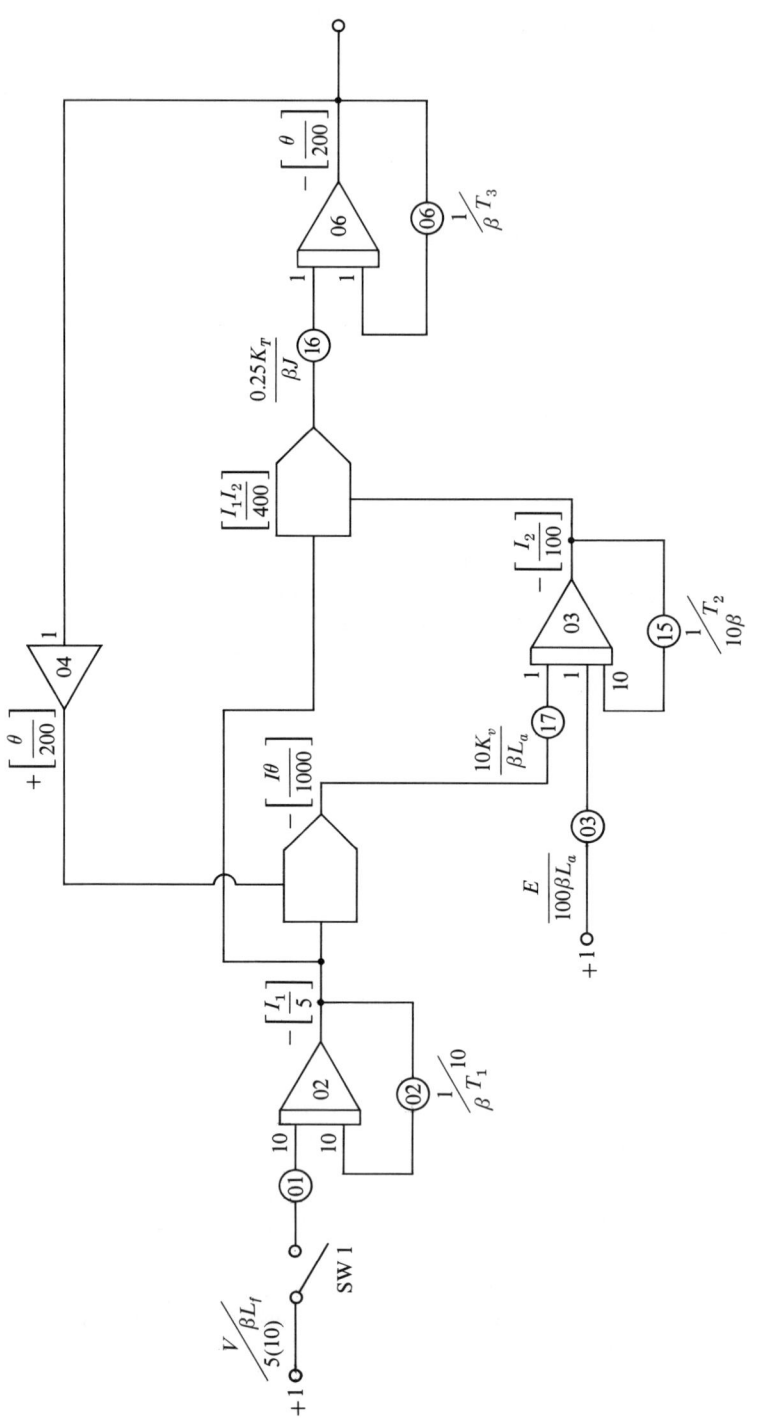

Fig. E3. Scaled Computer Solution ($\beta = 1$)

APPENDIX E. D.C. SERVO SIMULATION

SOLUTION TO PROBLEM STATEMENT III

The performance of a complete Static Check involves two parts: a *program check* and a *circuit check*. The solution presented here will depict the program check. The circuit check involves patching the program on the computer and measuring voltages to compare to calculations.

For a complete *program check*, calculations on the computer circuit itself must be made in addition to calculations from the original unscaled equations.

(a) Calculations based on the program can be performed directly on a copy of the circuit diagram if the outputs of all integrators and all pot settings are known. The integrator outputs are

$$-\left[\frac{I_1}{5}\right], \quad -\left[\frac{I_2}{100}\right], \quad +\left[\frac{\theta}{200}\right]$$

and their values are found from the values assumed for I_1, I_2 and θ in the statement of the problem; pot settings are calculated from the assumed values of the parameters (see problem statement). Figure E4 shows the diagram with these quantities marked in the appropriate place.

Now, without any other information except that which is shown in Fig. E4, all amplifier outputs can be calculated. Also, all derivative inputs of integrators can be found. For example,

$$A02 = -1.00$$
$$A00 = -A02 = +1.00$$
$$A03 = -0.5$$
$$A05 = +A03 = +0.5$$
$$A06 = -0.5$$
$$A31 = +0.5$$
$$A30 = -A31 = -0.5$$
$$A13 = -/A31//A02/ = -0.5$$
$$A25 = +/A03//A02/ = +0.5$$

with switch SW *I* closed:

$$D02 = 10(P01) + [1] + 10(P02)[A02]$$
$$= 10(0.4) + [1] + 10(1.0) - [1]$$
$$= -6.00$$
$$D03 = 1(P17)[A13] + 1(P03) + [1] + 10(P15)[A03]$$
$$= 1(0.0667) - [0.5] + 1(0.222) + [1] + 10(0.25)$$
$$- [0.5] = -1.06135$$
$$D06 = 1(P16)[A25] + 1(P06)[A06]$$
$$= 1(0.25) + [0.5] + 1(0.2) - [0.5]$$
$$= +0.025.$$

The calculated values for Static Test are shown in Fig. E5.

(b) Calculations based on the original problem are performed for amplifier outputs and derivative values as follows: The output of each

180 ANALOG/LOGIC COMPUTER PROGRAMMING AND SIMULATION

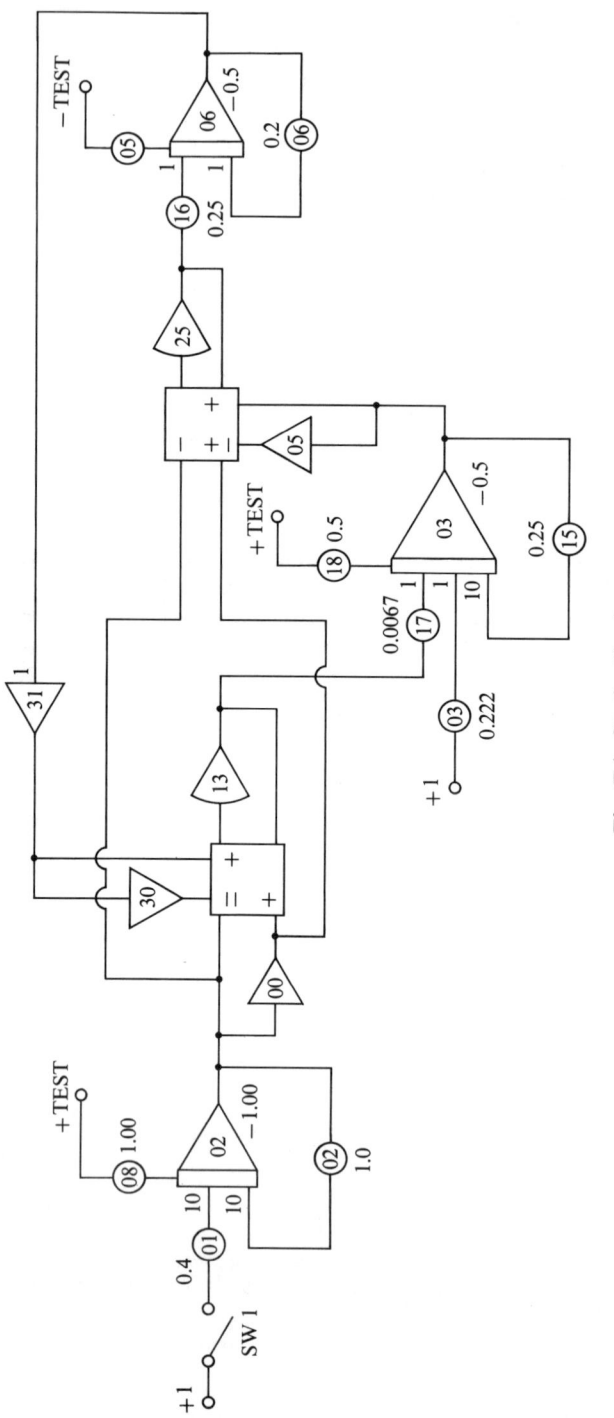

Fig. E4. Static Check

APPENDIX E. D.C. SERVO SIMULATION 181

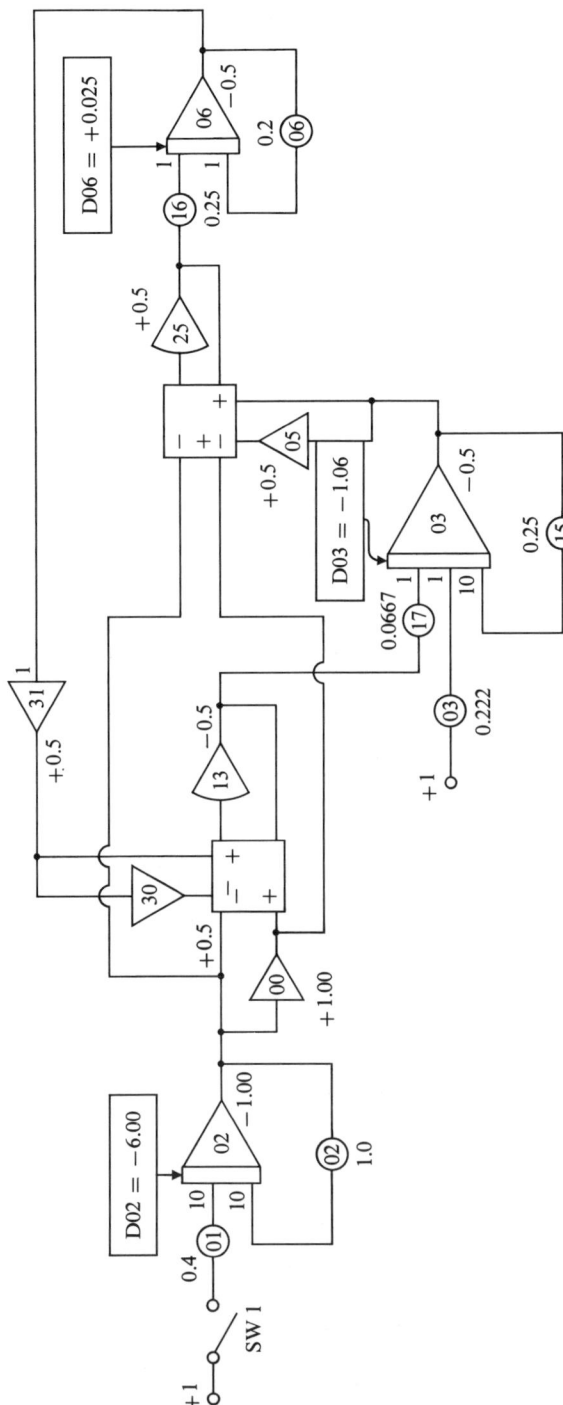

Fig. E5. Complete Static Check

amplifier is known in terms of the problem variables and a scale factor. Knowing the assumed values of these allows the calculations of these outputs without referring to the diagram at all. The scaling table lists the computer variables (i.e., the amplifier outputs) as

$$\left[\frac{I_1}{5}\right]$$

$$\left[\frac{I_2}{100}\right]$$

$$\left[\frac{\theta}{200}\right]$$

and values for I_1, I_2 and θ are known from the problem statement. Thus,

$$\text{A02} = -\left[\frac{I_1}{5}\right] = -\left[\frac{5}{5}\right] = -1.00$$

$$\text{A03} = -\left[\frac{I_2}{100}\right] = -\left[\frac{50}{100}\right] = -0.500$$

$$\text{A06} = +\left[\frac{\theta}{200}\right] = \left[\frac{100}{200}\right] = +0.500$$

$$\text{D02} = -\frac{d}{d\tau} - \left[\frac{I_1}{5}\right]$$

$$\frac{d}{d\tau}\left[\frac{I_1}{5}\right] = 10\left(\frac{V}{10(5)\beta L_f}\right) - 10\left(\frac{1}{T_1\beta 10}\right)\left[\frac{I_1}{5}\right]$$

$$\frac{d}{d\tau}\left[\frac{I_1}{5}\right] = \frac{100}{5(5)} - (10)(+1) = -6.00$$

$$\text{D02} = -6.00$$

$$\text{D03} = -\frac{d}{d\tau} - \left[\frac{I_2}{100}\right] = \frac{d}{d\tau}\left[\frac{I_2}{100}\right]$$

$$\frac{d}{d\tau}\left[\frac{I_2}{100}\right] = 1\left(\frac{E}{100\beta L_a}\right) - 10\left(\frac{1}{T_2 10\beta}\right)\left[\frac{I_2}{100}\right]$$

$$- 1\left(\frac{10K_v}{\beta L_a}\right)\left[\frac{I_1\theta}{1000}\right]$$

$$\frac{d}{d\tau}\left[\frac{I_2}{100}\right] = \frac{200}{(100)(9)} - 2.5(+0.5) - \frac{10(0.06)}{9}\frac{5(100)}{1000}$$

$$\frac{d}{d\tau}\left[\frac{I_2}{100}\right] = -1.06135$$

$$\text{D03} = -1.06135$$

$$\text{D06} = -\frac{d}{d\tau}\left[-\frac{\theta}{200}\right] = \frac{d}{d\tau}\left[\frac{\theta}{200}\right]$$

APPENDIX E. D.C. SERVO SIMULATION

$$\frac{d}{d\tau}\left[\frac{\theta}{200}\right] = 1\left(\frac{2.5K_T}{\beta J}\right)\left[\frac{I_1 I_2}{500}\right] - 1\left(\frac{1}{T_3\beta}\right)\left[\frac{\theta}{200}\right]$$

$$\frac{d}{d\tau}\left[\frac{\theta}{200}\right] = 1\frac{(2.5)(0.5)}{5}\left[\frac{5(50)}{500}\right] - (0.2)\left(\frac{100}{200}\right)$$

$$\frac{d}{d\tau}\left[\frac{\theta}{200}\right] = +0.025$$

$$D06 = +0.025.$$

Comparisons are now in order. Checking the values calculated in (b) with the ones calculated on the diagram in (a) shows 100% agreement. Thus, it is safe to say that the program of Fig. E3 truly represents the problem described in Part I. These values may now be entered on pot and amplifier sheets.

(d) The final effort in checking involves patching the program, setting pots, establishing the selected integrator outputs, and measuring all amplifier outputs and derivatives. The measured results should agree with calculated values to better than 1% or an error is indicated.

Proper integrator outputs for checking this problem are assured by the addition of test IC's for the three integrators through pots 18, 00 and 05 shown in Fig. E6.

184 ANALOG/LOGIC COMPUTER PROGRAMMING AND SIMULATION

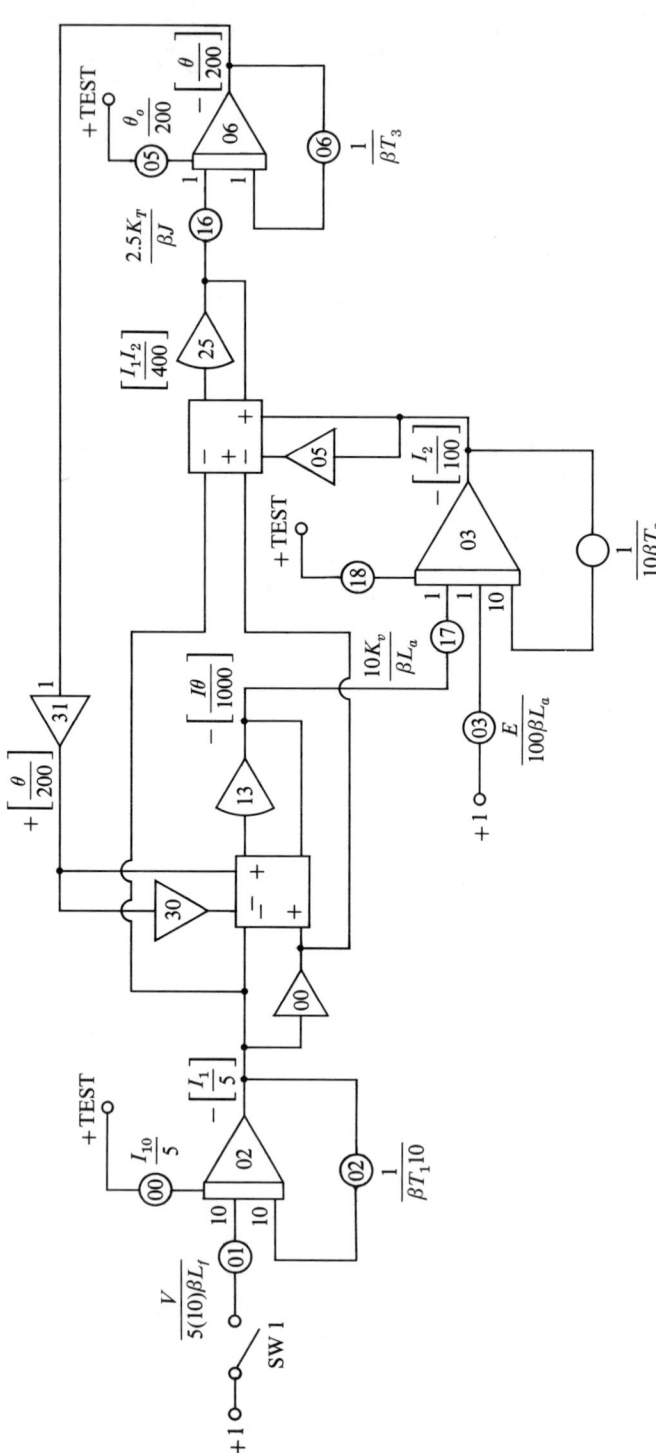

Fig. E6. Final Computer Program

APPENDIX E. D.C. SERVO SIMULATION

AMPLIFIER ASSIGNMENT SHEET

Date _____   Problem ___d.c. Servo Motor___

Amp. No.	FB	Output Variable	Static Check Calculated Check Pt.	Static Check Calculated Output	Static Check Measured Check Pt.	Static Check Measured Output	Notes
00	inv.	$+I_{1/5}$		$+1.00$			
01							
02	$\int$	$-I_{1/5}$	$+0.600*$	$+1.00$			
03	$\int$	$-I_{2/100}$	$+0.106*$	-0.500			*Measured through Check Amp. of Gain − 1/10
04							
05	inv.	$+I_{2/100}$		$+0.500$			
06	$\int$	$-\dfrac{\theta}{200}$	-0.025	-0.500			
07							
08							
09							
10							
11							
12							
13	H.G.	$-\dfrac{I\theta}{1000}$	-0.500				
14							
25	H.G.	$I_1 I_{2/400}$	$+0.500$				
26							
27							
28							
29							
30	inv.	$-\dfrac{\theta}{200}$	-0.500				
31	inv.	$+\dfrac{\theta}{200}$	$+0.500$				

POTENTIOMETER ASSIGNMENT SHEET

Date _____   Problem ___d.c. Servo Motor___

Pot. No.	Parameter Description	Setting Static Check	Static Check Output Voltage	Setting Run Number 1	///// Setting Run 2	Pot. No.
00	$I_{10/5}$	1.00		0.2203	—	
01	$V/5(10)\beta L_f$	0.400		0.800	Variable in 5 Steps	
02	$I/\beta T_1 10$	1.00		1.00	—	
03	$E/100\beta L_a$	0.222		0.500	—	
04						
05	$\theta_o/200$	0.500		0.799	—	
06	$I/\beta T_3$	0.200		0.200	—	
07						
08						
09						
10						
11						
12						
13						
14						
15	$1/10\beta T_2$	0.25		0.25	—	
16	$2.5 K_{T/\beta} J$	0.25		0.25	—	
17	$10 K_{v/\beta} L_a$	0.0667		0.0667	—	
18						
19						
20						
21						
22						
23						
24						

APPENDIX F

A CO_2 Rebreathing System

Introduction

The CO_2 partial pressure in a human being's mixed venous blood can be determined by measuring the CO_2 partial pressure in a bag, into which the subject is breathing, exhaling and rebreathing. The following simulation of the "bag rebreather" depends on the assumptions that

1. the ideal gas law holds, for constant volume and temperature conditions,
2. the arterial blood leaving the alveolus is in diffusion equilibrium with the alveolar gas, and
3. the same equilibrium condition exists for venous blood leaving the tissues.

An extension of this problem may be used in the study of air-pollution effects.

Mathematical Model

The equations describing the action of the system and typical parameter values are

$$\dot{P}_B = \frac{Q_A}{V_B}(P_A - P_B) \tag{F1}$$

$$\dot{P}_A = \frac{Q_A}{V_A}(P_B - P_A) + \frac{Q_H S(P_o - 47)}{V_A}(P_V - P_A) \tag{F2}$$

$$\dot{P}_V = \frac{Q_H S(P_o - 47)}{V_T}(P_A - P_V) + \frac{Q_o}{V_T}(P_o - 47) \tag{F3}$$

where

P_A = Alveolar CO_2 partial pressure, mm Hg
P_B = Bag CO_2 partial pressure, mm Hg
P_V = Venous CO_2 partial pressure, mm Hg
P_o = Barometric pressure, 760 mm Hg
Q_A = Alveolar ventilation, 8 liters/min.
Q_H = Cardiac output, 5 L/min.
Q_o = CO_2 output, 1/4 L/min.
V_A = Alveolar equivalent volume, 6L
V_B = Bag volume, 3L
V_T = Tissue equivalent volume, 75L
S = a constant, 0.0064 L/L/mm Hg
47 = Vapor pressure of water at body temperature, mm Hg.

Problem Statement

Program the analog computer to simulate this rebreathing system with a minimum number of amplifiers. Assume that none of the partial pressures exceeds 100 mm Hg. Suppose that $P_B(0) = 57$ mm Hg and use the following for 3 runs

Run	1	2	3	
$P_A(0)$	20	40	80	mm Hg
$P_V(0)$	24	46	88	mm Hg.

Run 2 allows the investigation of system response with near-normal partial pressure levels, while Run 3 involves higher-than-normal and Run 1, lower-than-normal partial pressures.

Problem Solution—Part 1

Unscaled Diagram

Since the parameters of interest in this problem are the initial conditions, and these are always isolated, the best program choice would therefore be the one using fewest amplifiers. To make the equations easier to handle, define: $Q_H S(P_o - 47) = K$. Then the equations are:

$$-\dot{P}_B = -\left(\frac{Q_A}{V_B}\right)P_A + \left(\frac{Q_A}{V_B}\right)P_B \tag{F4}$$

APPENDIX F. A CO_2 REBREATHING SYSTEM

$$\dot{P}_A = \left(\frac{Q_A}{V_A}\right)P_B + \left(\frac{K}{V_A}\right)P_V - \left(\frac{Q_A + K}{V_A}\right)P_A \quad \text{(F5)}$$

$$-\dot{P}_V = +\left(\frac{K}{V_T}\right)P_V - \left(\frac{K}{V_T}\right)P_A - \frac{Q_o(P_o - 47)}{V_T}. \quad \text{(F6)}$$

A program for these equations is shown in Fig. F1.

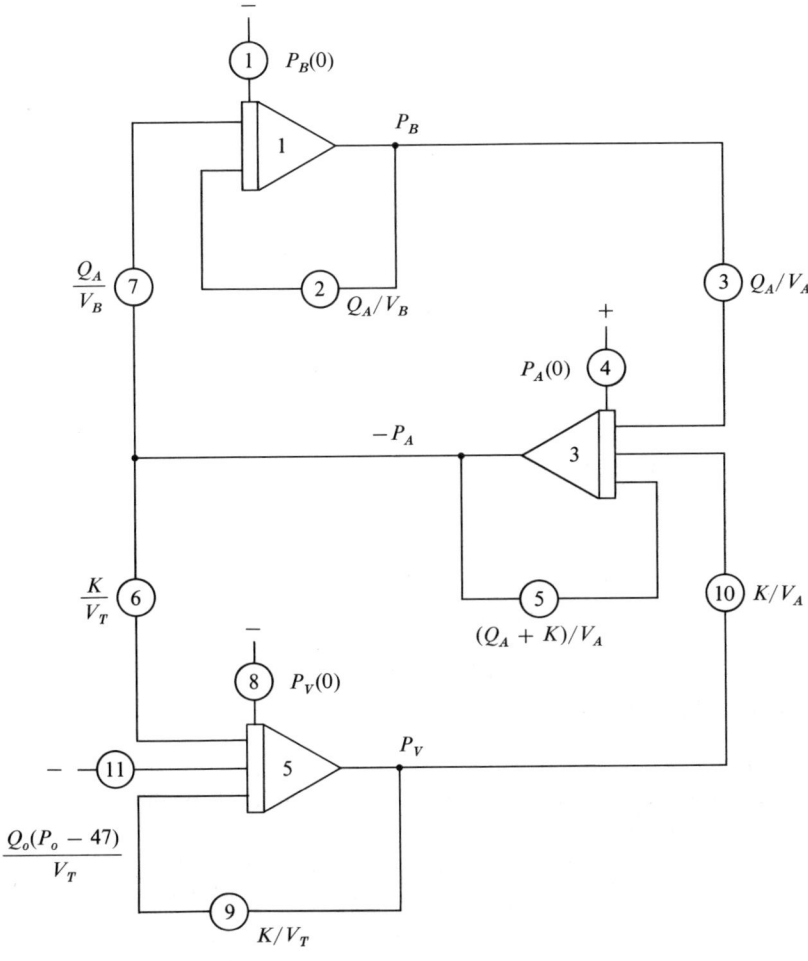

Fig. F1. Unscaled Diagram

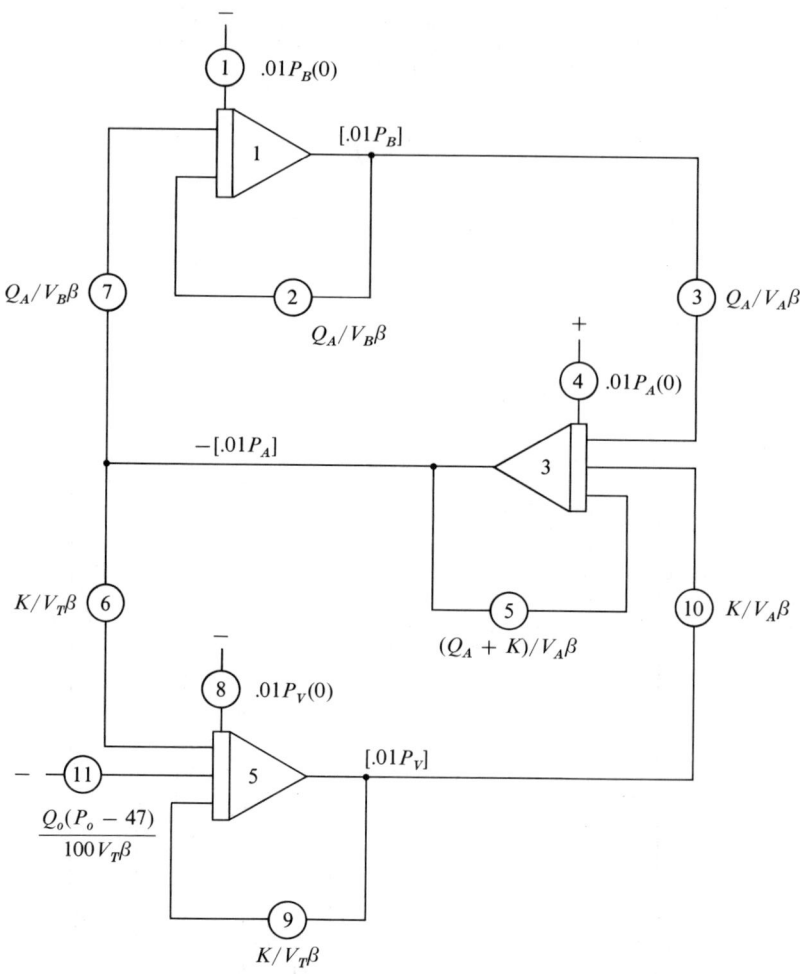

Fig. F2. Scaled Diagram

PROBLEM SOLUTION—PART 2

Scaling

The scaling table below is a result of values given in problem statement.

Physical Variable	Maximum	Computer Variable
P_B	100	$[.01 P_B]$
P_A	100	$[.01 P_A]$
P_V	100	$[.01 P_V]$

APPENDIX F. A CO_2 REBREATHING SYSTEM

The scaled equations are then quite simple.

$$-\frac{d}{d\tau}[.01P_B] = -\left(\frac{Q_A}{\beta V_B}\right)[.01P_A] + \left(\frac{Q_A}{\beta V_B}\right)[.01P_B] \quad \text{(F7)}$$

$$\frac{d}{d\tau}[.01P_A] = \left(\frac{Q_A}{\beta V_B}\right)[.01P_B] + \left(\frac{K}{\beta V_A}\right)[.01P_V]$$
$$- \left(\frac{Q_A + K}{\beta V_A}\right)[.01P_A] \quad \text{(F8)}$$

$$-\frac{d}{d\tau}[.01P_V] = \left(\frac{Q_o(P_o - 47)}{100 V_T \beta}\right)[-1] - \left(\frac{K}{\beta V_T}\right)[.01P_A]$$
$$+ \left(\frac{K}{\beta V_T}\right)[.01P_V] \quad \text{(F9)}$$

The value of β is chosen by inspection of the integrator gains. They are:

$$Q_A/V_B = 2.667 \text{ (minute)}^{-1}$$
$$Q_A/V_A = 1.333 \text{ (minute)}^{-1}$$
$$K/V_A = 3.803 \text{ (minute)}^{-1}$$
$$K/V_T = .3042 \text{ (minute)}^{-1}$$
$$(Q_A + K)/V_A = 5.136 \text{ (minute)}^{-1}$$
$$Q_o(P_o - 47)/100 V_T = .02377 \text{ (minute)}^{-1}.$$

Excepting $Q_o(P_o - 47)/100 V_T$, all gains fit into the range 0.3042 to 5.136, which fits nicely into our rule-of-thumb range of 0.1000 to 10.000. Hence, $\beta = 1$. Note that there is an implied factor of 60 (1 second of computer time = 1 minute of "rebreather" time) because of the problem's unit of time. Finally, the gains must be separated into input factors and pot settings. The results are shown in Fig. F3.

PROBLEM SOLUTION—PART 3

Static Check

Since all integrators have non-zero IC's and all pots have non-zero settings, then any of the three runs provide reasonable static check values. The parameters and their values are:

$$P_o = 760, \ Q_A = 8, \ Q_H = 0.25$$
$$V_A = 6, \ V_B = 3, \ V_T = 75, \ S = 6.4 \times 10^{-3}$$
$$K = 22.82, \ \beta = 1, \ P_B(0) = 57.$$

Note that the outputs of amplifiers 1, 2, and 3 are the selected IC's and the outputs of 2, 4, and 6 are the respective inverted values. Hence, the derivative values are of major interest.

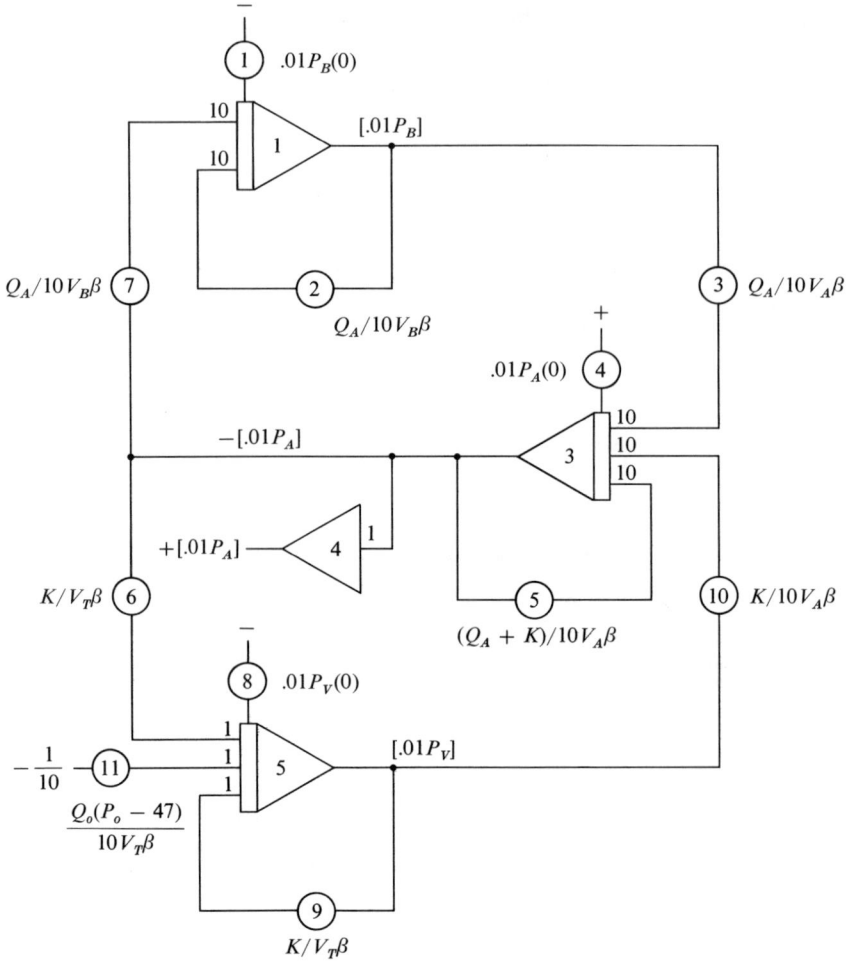

Fig. F3. Final Scaled Diagram

Run 1 Values: $P_A(0) = 20$, $P_V(0) = 24$.

$$-\dot{P}_B = (Q_A/V_B)(P_B - P_A)$$
$$= (8/3)(57 - 20)$$
$$= 98.67$$

$$\dot{P}_A = (Q_A/V_A)P_B + (K/V_A)P_V - \left(\frac{Q_A + K}{V_A}\right)P_A$$
$$= (8/6)(57) + \left(\frac{22.82}{6}\right)(24) - \left(\frac{30.82}{6}\right)(20)$$
$$= 64.56$$

$$-\dot{P}_V = (K/V_T)P_V - (K/V_T)P_A - \frac{Q_o(P_o - 47)}{V_T}$$

$$= \left(\frac{22.82}{75}\right)(24) - \left(\frac{22.82}{75}\right)(20) - \frac{(.25)(713)}{75}$$

$$= -1.16.$$

The derivative values can be calculated from the diagram by observing that the given parameter values are equivalent to $.01P_B = 0.570$, $.01P_A = 0.200$, $.01P_V = 0.240$ as the outputs of amplifiers 1, 2, and 3, while the pot settings are

$P2 = P7 = .2667$
$P3 = .1333$
$P5 = .5136$
$P6 = P9 = .3042$
$P10 = .3803$
$P11 = .2377.$

Hence,

$$D1 = 10(P2)(A1) + 10(P7)(A3)$$
$$= 10(.2667)(.570) + 10(.2667)(-.200)$$
$$= 0.9867.$$

But,

$$D1 = -\frac{1}{\beta}\frac{d}{dt}[.01P_B].$$

Thus,

$$-\dot{P}_B = +98.67 \text{ which agrees with the calculation given previously.}$$

Similarly,

$$D3 = 10(P3)(A1) + 10(.3803)(+.240) + 10(.5136)(-.200)$$
$$= 10(.1333)(+.570) + 10(.3803)(+.240)$$
$$\quad + 10(.5136)(-.200) = -0.6456$$

$$= +\frac{1}{\beta}\frac{d}{dt}[.01P_A].$$

Thus,

$$\dot{P}_A = +64.56, \text{ also from the previous calculation.}$$

Finally,

$$D5 = (1)(P6)(A3) + (1)(P11)\left(-\frac{1}{10}\right) + 1(P9)(A5)$$

$$= (.3042)(-.200) + (.2377)\left(-\frac{1}{10}\right) + (.3042)(.240)$$

$$= -0.0116.$$

Now, this is too small a value to measure accurately, so a change in our original choice is now dictated.

Let's try changing $P_A(0)$ to 50. Then

$$-\dot{P}_B = (Q_A/V_B)(P_B - P_A) = (8/3)(57 - 50) = 18.67$$

$$\dot{P}_A = (Q_A/V_A)P_B + (K/V_A)P_V - \left(\frac{Q_A + K}{V_A}\right)P_A$$

$$= (8/6)(57) + \left(\frac{22.82}{6}\right)(24) - \left(\frac{30.82}{6}\right)(50)$$

$$= -88.50$$

$$-\dot{P}_V = (K/V_T)(P_V - P_A) - \frac{Q_o(P_o - 47)}{V_T}$$

$$= \left(\frac{22.82}{75}\right)(24 - 50) - \frac{.25(713)}{75}$$

$$= -10.288.$$

Turning to the diagram in Fig. F4, we see that P4 has been changed which, in turn, changes A3, A4, D1, D3, and D5. The new values are

P4 = .5000
A3 = .5000
D1 = 10(P2)(A1) + 10(P7)(A3) = .1867
D3 = 10(P3)(A1) + 10(P10)(A5) + 10(P5)(A3) = −.8950

and

D5 = 1(P6)(A3) + 1(P11)(−.1) + 1(P9)(A5) = −.1029.

Since $\beta = 1$, D1 = $-\dot{P}_B/100$, D3 = $+\dot{P}_A/100$ and D5 = $-\dot{P}_V/100$, we have successfully completed the paper portion of the Static Check.

APPENDIX F. A CO_2 REBREATHING SYSTEM

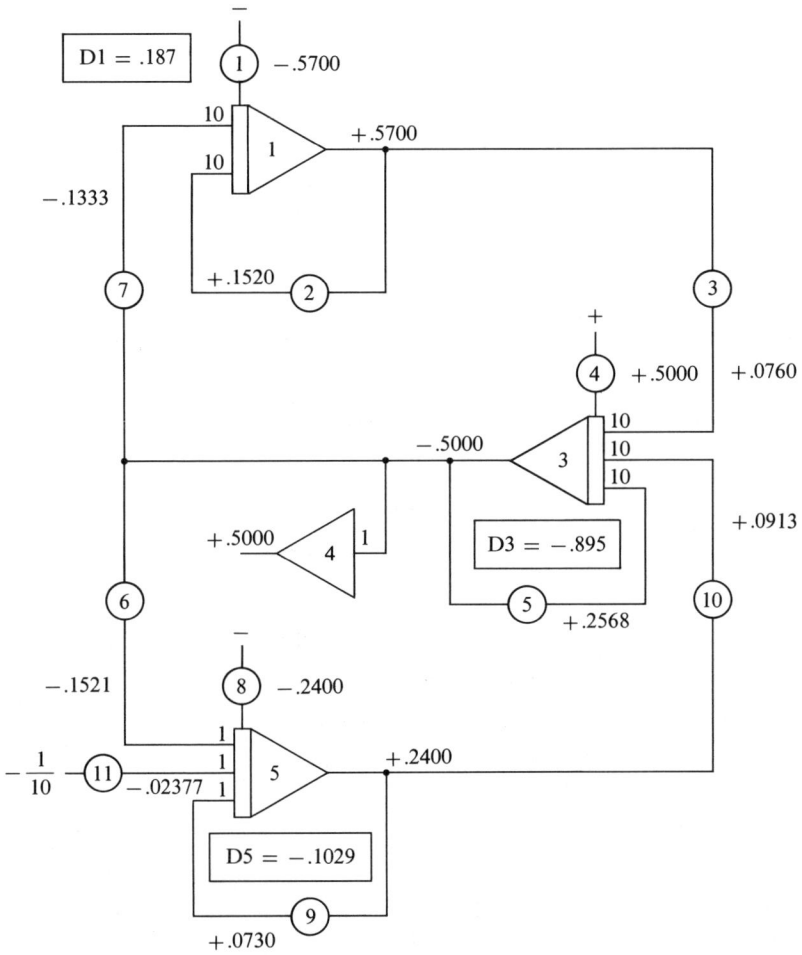

Fig. F4. Voltage Check ($P_A = 50$)

POTENTIOMETER ASSIGNMENT SHEET

Date _____   Problem ___CO_2 Rebreather___

Pot. No.	Parameter Description	Setting Static Check	Static Check Output Voltage	Setting Run Number 1	Notes	Pot. No.
00					$\beta = 1$	
01	$.01 P_B(0)$	.5700	→		Parameter	
02	$Q_A/10 V_B \beta$	.2667	→			
03	$Q_A/10 V_A \beta$	.1333	→			
04	$.01 P_A(0)$	.5000		.2000	Parameter	
05	$(Q_A + K)/10 V_A \beta$	.5136	→			
06	$K/V_T \beta$	.3042	→			
07	$Q_A/10 V_B \beta$	.2667	→			
08	$.01 P_V(0)$	.2400	→		Parameter	
09	$K/V_T \beta$	.3042	→			
10	$K/10 V_A \beta$	.3803	→			
11	$Q_o(P_o - 47)/10 V_T \beta$	.2377	→			
12						
13						
14						
15						
16						
17						
18						
19						
20						
21						
22						
23						
24						

AMPLIFIER ASSIGNMENT SHEET

Date _____   Problem ___CO_2 Rebreather___

Amp. No.	FB	Output Variable	Static Check				Notes
			Calculated		Measured		
			Check Pt.	Output	Check Pt.	Output	
00							
01	∫	$.01P_B$	+.1867	.5700			Check Amplifier Gain = −1
02							
03	∫	$-.01P_A$	−.8950	−.5000			
04	−1	$.01P_A$		+.5000			
05	∫	$.01P_V$	−.1029	.2400			
06							
07							
08							
09							
10							
11							
12							
13							
14							
15							
16							
17							
18							
19							
20							

APPENDIX G

Analysis of Kinetics Data

Introduction

The rate at which a chemical reaction occurs is of considerable interest to chemical engineers. This problem is intended to demonstrate the use of the analog computer in finding the best fit for experimental data obtained in a reaction.

PROBLEM STATEMENT I

Data obtained in a bench-scale study of a chemical reaction are as follows:

Time Minutes	Normalized Concentration
0	1.000
1	.888
2	.808
3	.728
4	.660
5	.616
6	.572
7	.536
8	.500
9	.472
10	.444
15	.348
20	.285
25	.242
30	.211
40	.166
50	.138

As a first approximation, it can be assumed that the reaction can be described by the differential equation

$$\frac{dC}{dt} = -KC^n.$$

Devise a computer circuit to find the rate constant K and order n of the reaction to give the best fit to the experimental data. Use exactly 7 amplifiers. Prepare the simulation to plot C versus t.

SOLUTION TO PROBLEM STATEMENT I

The direct programming of the equation

$$\frac{dC}{dt} = -KC^n \tag{G1}$$

appears difficult because n is an exponent and it need not be an integer. In order to solve the problem efficiently, we need an equation where n is not an exponent. The general procedure in a case like this is to take the derivative with respect to time of the equation and produce another differential equation that can be programmed more easily.

Differentiating Eq. (G1) with respect to time:

$$\ddot{C} = -KnC^{n-1}\dot{C}$$

$$\ddot{C} = -Kn\frac{C^n}{C}\dot{C}$$

but, from (G1),

$$KC^n = -\dot{C}$$

$$\ddot{C} = n\frac{(\dot{C})^2}{C}. \tag{G2}$$

This equation can be programmed easily. Figure G1 indicates the unscaled computer solution.

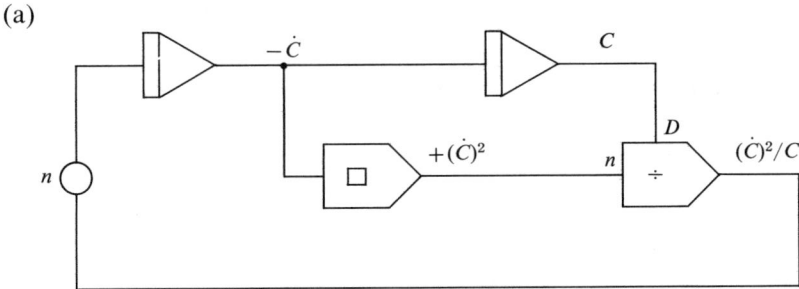

Fig. G1. Block Diagram

200 ANALOG/LOGIC COMPUTER PROGRAMMING AND SIMULATION

(b)

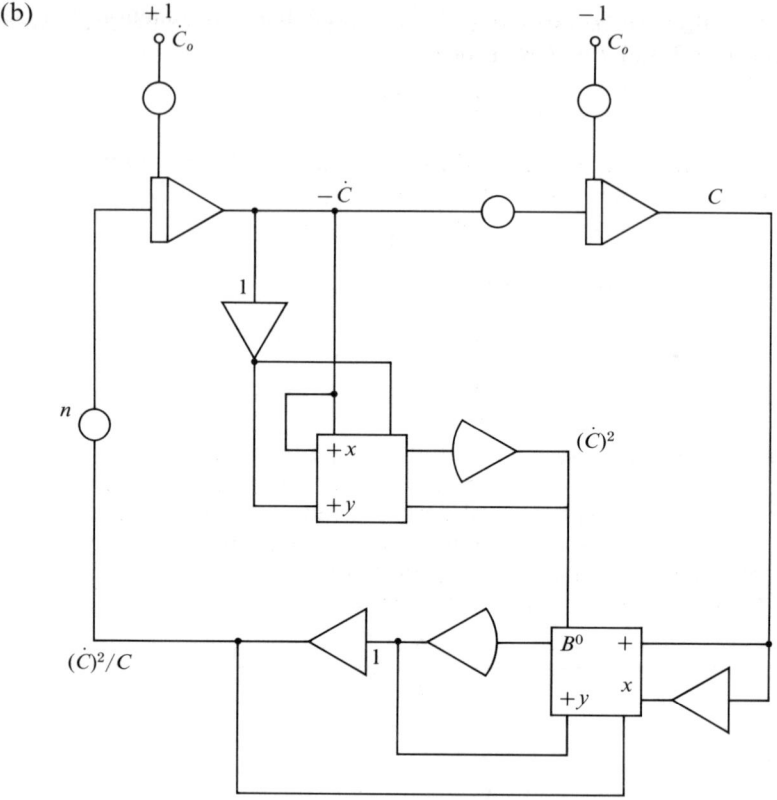

Fig. G1. Unscaled Computer Diagram Showing All Amplifiers (See Computer Reference Handbook for Patching Details)

PROBLEM STATEMENT II

The Kinetics problem was programmed above without scaling; now the problem should be amplitude and time scaled.

The maximum values of the variables should be estimated from the data presented in Part I.

From experience, it can be estimated that n will probably be between 1 and 4.

SOLUTION TO PROBLEM STATEMENT II

The maximum value of C is given in the data as 1.00. Since we are considering an exponential fit, the maximum value of $\dot{C}$ will be its initial

APPENDIX G. ANALYSIS OF KINETICS DATA

value, and this can be estimated as the slope of the concentration curve between $t = 0$ and $t = 1$ min. or

$$|\dot C \max| = \frac{1.00 - 0.888}{1 \text{ min.}} = 0.112/\text{min.}$$

The maximum value of $(\dot C)^2/C$ is the maximum of $\dot C$ over the minimum of C. For

$$\dot C \max. = 0.112/\text{min.}$$
$$C \min. = 0.1$$
$$\frac{(\dot C \max.)^2}{C} = \frac{(0.112)^2}{0.1} = .125.$$

Hence we can use the dividing circuit's built-in scale factor.

It is now necessary to make up a table with the computer variables including the outputs of all amplifiers in the list.

Physical Variable	Rounded-up Maximum	Computer Variable
C	1	$[C]$
$\dot C$	0.5	$[2\dot C]$

The computer variables are now substituted in Eq. (G2) to obtain the scaled equations.

$$\frac{d}{dt}\frac{[2\dot C]}{2} = n\left[\frac{[2\dot C]^2}{[C]}\right]1/4$$

$$\frac{d}{dt}[2\dot C] = \left[\frac{[2\dot C][2\dot C]}{[C]}\right](n/2) \tag{G3}$$

Equation (G3) is the properly scaled equation; however, one must look at the integrator gains to be certain that they are between 0.1 and 10. A $\beta = 1$ will make the gains and pot settings reasonable. This is consistent with the original data. The study was taken over a 50 minute period; that is, for $\beta = 1$, 50 minutes of problem time is equivalent to 50 seconds of computer time.

The completely scaled equation is indicated in Eq. (G4).

$$\frac{d}{d\tau}[2C] = 10\frac{n}{20\beta}\frac{[2\dot C][2\dot C]}{[C]} \tag{G4}$$

The completely scaled computer diagram is shown in Fig. G2.

To obtain a complete solution to this problem one must fit the computer solution to the experimentally obtained data. Basically, there are two ways of doing this:

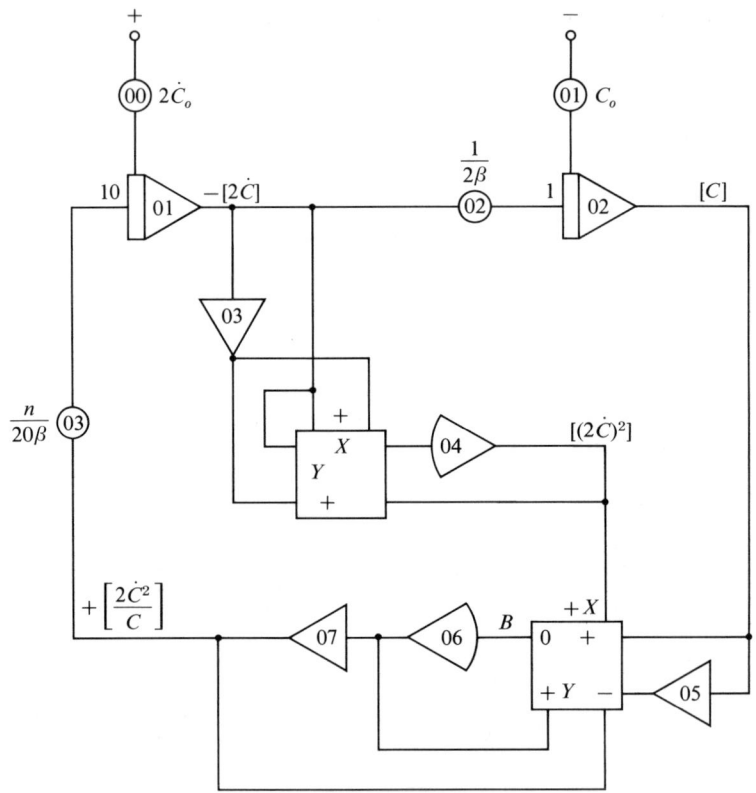

Fig. G2. Scaled Computer Diagram ($\beta = 1$)

One method is to plot the concentration versus time, and using this curve on the X-Y plotting board, vary pots P00 and P03 until the computer solution matches the plotted curve.

An alternate attack uses the repetitive operation feature. The given data are placed on a function generator and viewed on a Rep-Op Display Scope. The computer solution is also placed on the scope, and the pots varied until the two curves match. An accurate trace of the computer solution can then be plotted in a slow time.

The value of K is determined from the setting of P00, the value of n from P03.

An example is shown below.

$$\frac{n}{20\beta} = \text{P03}$$

$$n = 20(\text{P03})$$

since
$$\beta = 1.$$
Also,
$$2\dot{C}_o = \text{P00}$$
and
$$\dot{C}_o = -KC_o{}^n.$$
Therefore,
$$K = \left(\frac{\text{P00}}{2C_o{}^n}\right).$$

Since $C_o = 1$, the calculation is easy, and $K = (\text{P00}/2)$.

Problem Statement III

In the process of programming, scaling and patching a problem, numerous opportunities for error arise. Since even one error in a program can invalidate the entire result, it is absolutely necessary to detect and correct these errors before actual computation starts. The method of detecting and correcting programs is called Static Check.

Perform a Static Check for the Analysis of Kinetic Data problem where

$$C = 1$$
$$\dot{C} = 400 \times 10^{-3}/\text{min}.$$
$$n = 3.$$

After the Static Check is completed, patch the problem and attempt to find the values of n and K to fit the data given in Part I.

Solution to Problem Statement III

The performance of a complete Static Check involves two parts: a *program check* and a *circuit check*. The solution presented here will depict the program check. The circuit check involves patching the program on the computer and measuring voltages to compare to calculations.

For a complete *program check*, calculations on the computer circuit itself must be made in addition to calculations from the original unscaled equations.

(a) Calculations based on the program can be performed directly on a copy of the circuit diagram if the outputs of all integrators and all pot settings are known. The integrator outputs are

$$-[2\dot{C}], \quad +[C]$$

and their values are found from the values assumed for $\dot{C}$ and C in the statement of the problem; pot settings are calculated from the assumed values of the parameters (see problem statement). Figure G3 shows the diagram with these quantities marked in appropriate places.

Now without any other information except that which is shown in

204 ANALOG/LOGIC COMPUTER PROGRAMMING AND SIMULATION

Fig. G3, all amplifier outputs can be calculated. Also, all derivative inputs of integrators can be found. For example,

$$A01 = -0.800$$
$$A02 = +1.00$$
$$A03 = -A01 = +0.800$$
$$A04 = +(0.8)^2 = +0.64$$
$$A05 = -A02 = -1.00$$
$$A06 = -0.64$$
$$A07 = -A06 = +0.64$$

and

$$D01 = -10(P03)[A07]$$
$$D01 = -10(0.15)[+0.64]$$
$$D01 = -0.960$$
$$D02 = -1(P02)[A01]$$
$$D02 = -1(0.5)[-0.800]$$
$$D02 = +0.400.$$

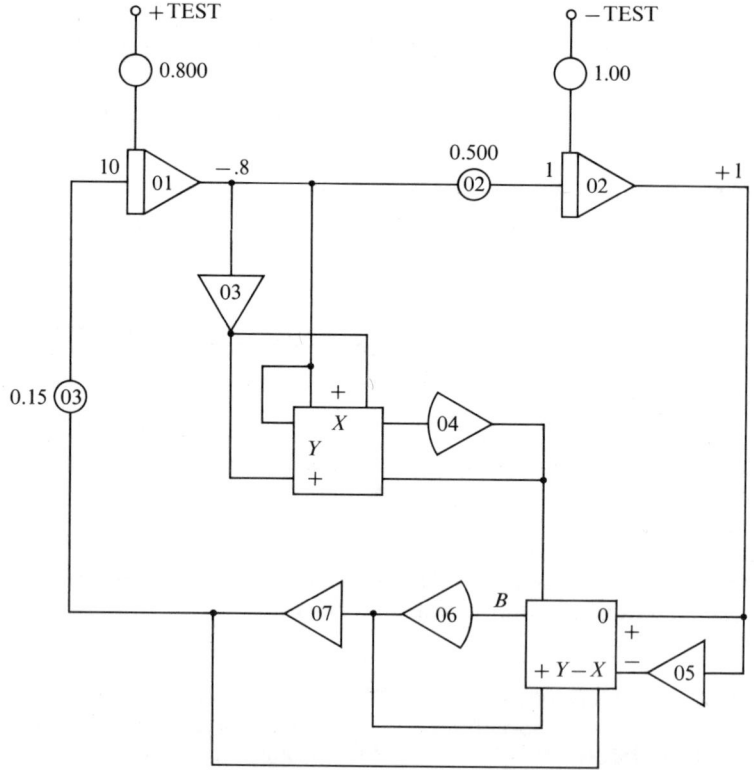

Fig. G3. Static Check

These values are shown in Fig. G4, the completed program check.

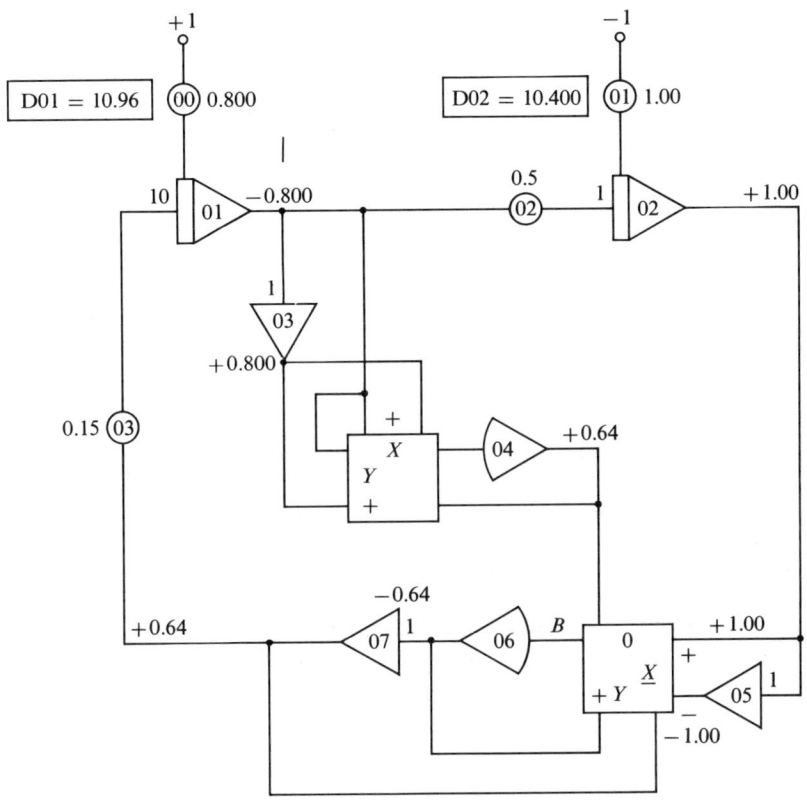

Fig. G4. Completed Static Check

(b) Calculations based on the original problem are performed for amplifier outputs and derivative values as follows:

$$A01 = -[2\dot{C}] = -[2(400 \times 10^{-3})] = -0.800$$
$$A02 = +[1] = +[1] = +1.00$$
$$A03 = +[2\dot{C}] = +0.800$$
$$A04 = +[2\dot{C}]^2 = -[(0.8)(0.8)] = -0.64$$
$$A05 = -[2\dot{C}]^2 = +0.64$$
$$A06 = -\frac{[2\dot{C}]^2}{[C]} = -0.64$$
$$A07 = +0.64$$

and

$$\text{D01} = +\frac{d}{d\tau}[-2\dot{C}] = -\frac{2\ddot{C}}{\beta}$$

where

$$\ddot{C} = n\frac{(\dot{C})^2}{C} = 3\frac{(0.4)^2}{1} = 0.480$$

n, C and $\dot{C}$ are the assumed values and $\beta = 1$. Then,

$$\text{D01} = -2\ddot{C} = -0.96$$

$$\text{D02} = +\frac{d}{d\tau}[C] = +\frac{\dot{C}}{\beta} = +0.400$$

$$\text{D02} = +0.400.$$

(c) Comparisons are now in order. Checking the values calculated in (b) with the ones calculated on the diagram in (a) shows 100% agreement. Thus, it is safe to say that the program of Fig. G2 truly represents the problem described in Part I. These values may now be entered on pot and amplifier sheets.

(d) The final effort in checking involves patching the program, setting pots, establishing the selected integrator outputs, and measuring all amplifier outputs and derivatives. The measured results should agree with the calculated values to better than 1% or an error is indicated.

Proper integrator outputs for checking this problem are assured by the addition of test IC's for the two integrators through pots 02 and 03 as shown in Fig. G5.

APPENDIX G. ANALYSIS OF KINETICS DATA

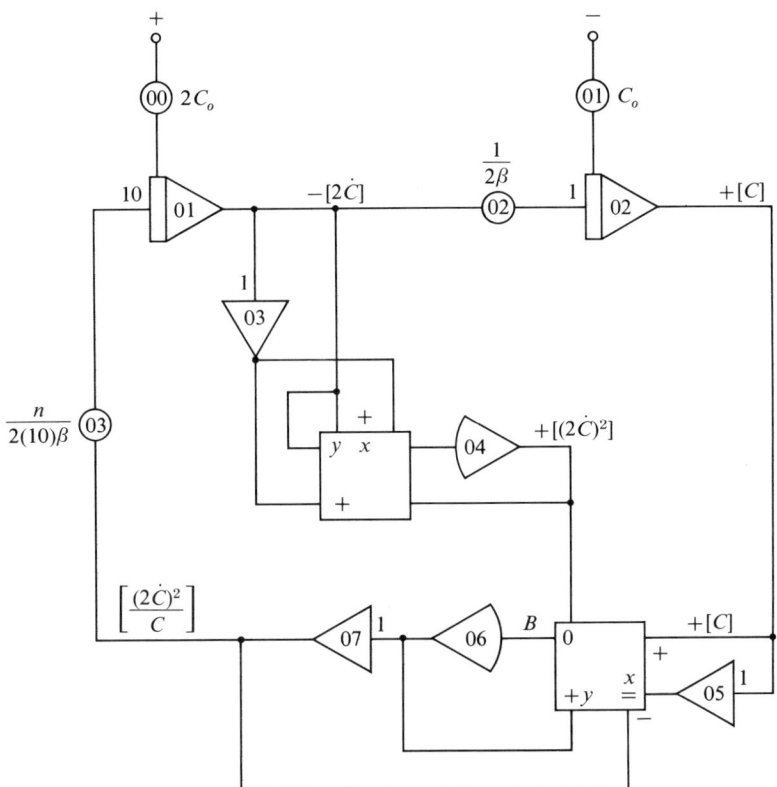

Fig. G5. Final Computer Diagram

POTENTIOMETER ASSIGNMENT SHEET

Date _____   Problem __Analysis of Kinetics Data__

Pot. No.	Parameter Description	Setting Static Check	Static Check Output Voltage	Setting Run Number 1	Notes	Pot. No.
00	$2C_o$ (1)	0.800		*	Parametric Var.	
01	C_o	1.000				
02	$1/2\beta$	0.500				
03	$n/20\beta$	0.150		*	Parametric Var.	
04						
05					$\beta = 1$	
06						
07						
08						
09						
10						
11					(1) $2\dot{C}_o = 2K$	
12						
13						
14						
15						
16						
17						
18						
19						
20						
21						
22						
23						
24						

APPENDIX G. ANALYSIS OF KINETICS DATA

AMPLIFIER ASSIGNMENT SHEET

Date _____   Problem __Analysis of Kinetics Data__

| Amp. No. | FB | Output Variable | Static Check |||| Notes |
| | | | Calculated || Measured || |
			Check Pt.	Output	Check Pt.	Output	
00		Deriv.					
01	∫	$-[2\dot{C}]$	-0.960	-0.800			
02	∫	$+[C]$	$+0.400$	$+1.000$			
03	-1	$+[2\dot{C}]$		$+0.800$			
04	M	$+[(2\dot{C})^2]$		$+0.640$			
05	-1	$-[2\dot{C}]$		-1.000			
06	D	$-[(2\dot{C})^2/C]$	—	-0.640			
07	-1	$+[(2\dot{C})^2/C]$		$+0.640$			
08							
09							
10							
11							
12							
13							
14							
15							
16							
17							
18							
19							
20							

APPENDIX H

Dynamic Behavior of Enzyme Systems

Introduction

The complex chemical reaction sequences that constitute the activity of living systems are the direct result of a class of biological catalysts known as enzymes. This demonstration note discusses the application of analog techniques in the study of the dynamic behavior of enzyme systems.

An enzyme can be defined as a long chain protein that changes the rate of a biochemical reaction but does not affect the nature of the final products. Enzymes thus behave much like ideal catalysts since they are effective in very small amounts and are usually unchanged in the course of reaction. Enzymes also exhibit typical catalytic specificity in their ability to accelerate chemical reactions and, like catalysts, do not change the nature of the equilibrium but instead change only the rate of attaining that equilibrium.

In homgeneous systems the reaction-rate (measured in amount of end product formed per unit time) also depends on the concentration of reactants. A second order reversible reaction may be written

$$A + B \underset{k_2}{\overset{k_1}{\rightleftharpoons}} C + D$$

in which k_1 and k_2 are known as specific rate constants. A differential equation giving the rate of product formation is then

$$\frac{dC}{dt} = k_1(AB)(B) - k_2(C)(D).$$

This general second order reaction and the catalytic role played by enzymes may be conveniently visualized with the aid of the energy diagram of Fig. H1. The formation of products C and D is preceded by the formation of a transient complex particle of higher potential energy. The energy difference between the initial state and the complex stage is known as the activation energy and is a measure of the difficulty with which the reaction takes place. Reactions with high activation energies proceed slowly at biologically normal temperatures since few molecules have sufficient thermal energy to overcome the energy barrier. In the catalyzed reaction, however, a new pathway of lower activation energy is provided and the reaction can take place at physiologically acceptable temperatures.

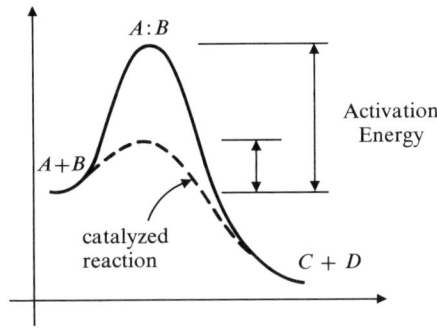

Fig. H1

As early as 1913 Michaelis-Menten postulated a similar mechanism to explain enzyme kinetics. It was assumed that when an enzyme catalyzes a specific reaction it first combines transiently with the *substrate* (substance on which the enzyme acts) to form the *enzyme-substrate complex*. A lock and key analogy has often been used to explain the mechanism since the substrate must conform in a very specific way to the enzyme active site for catalysis to occur. The active site then serves to modify the substrate molecule to provide a reaction path of lower activation energy.

The specificity of enzyme reactions are illustrated by the phenomenon of *competitive inhibition* shown diagrammatically in Fig. H2. As an example, consider the enzyme succinic dehydrogenase which catalyzes the oxidation of succinic acid. Normally the catalyzed reaction proceeds through the formation of an enzyme substrate-complex to the formation of reaction products and the reformation of free enzyme. But if malonic acid, whose structure closely resembles that of succinic acid, is added to the solution, the enzyme's effectiveness is considerably reduced. The malonic acid apparently competes for the active site that would normally be filled by succinic acid thus removing the enzyme from its catalytic role. Since substrate and inhibition compete for the same site, the degree of inhibition depends on the concentration ratio $[I]/[S]$ as well as the reaction rates involved in the formation of the $E:S$ and $E:I$ complexes.

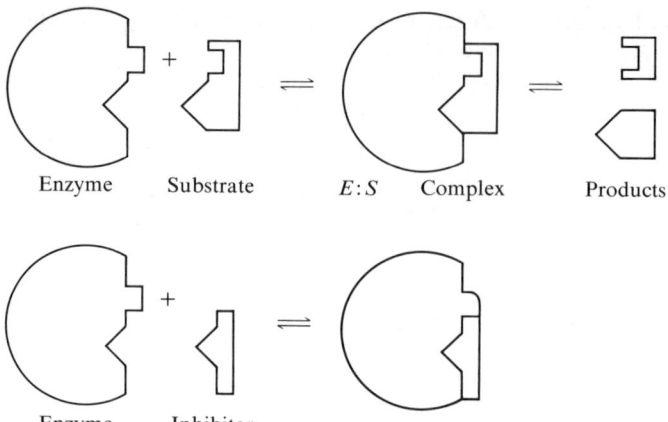

Fig. H2

Mathematical Formulation

The Michaelis-Menten hypothesis for the mechanism of enzyme action may be written

$$E + S \underset{k_2}{\overset{k_1}{\rightleftharpoons}} E:S \underset{k_4}{\overset{k_3}{\rightleftharpoons}} E + P. \tag{H1}$$

The phenomenon of competitive inhibition is included by providing reversible path for the formation of $E:I$ complex.

$$E + I \underset{k_6}{\overset{k_5}{\rightleftharpoons}} E:I \tag{H2}$$

The differential rate equations are then

$$\frac{dE}{dt} = -k_1[E][S] + k_2[E:S] + k_3[E:S] - k_4[E][P]$$
$$+ (k_5[E][I] - k_6[E:I]) \tag{H3}$$

$$\frac{dS}{dt} = -k_1[E][S] + k_2[E:S] \tag{H4}$$

$$\frac{d[E:S]}{dt} = k_1[E][S] - k_2[E:S] - k_3[E:S] + k_4[E][P] \tag{H5}$$

$$\frac{dP}{dt} = k_3[E:S] - k_4[E][P] \tag{H6}$$

$$\left(\frac{dE:I}{dt} = k_5[E][I] - k_6[E:I]\right) \tag{H7}$$

$$\left(\frac{dI}{dt} = k_5[E][I] - k_6[E:I]\right) \tag{H8}$$

APPENDIX H. DYNAMIC BEHAVIOR OF ENZYME SYSTEMS

where the contributions made by the action of inhibitors are enclosed in parentheses.

For reasons of mathematical simplicity, the standard biochemical treatment of enzyme kinetics ignores the k_4 process and assumes that $S = E$. This quasi steady-state assumption implies that complex concentration $[E:S]$ is nearly constant throughout the reaction and is only valid for low enzyme concentrations.

If we define E_o as the initial concentration of enzyme

$$[E] = [E_o] - [E:S]$$

the Eq. (H1) (without inhibitor) may be rewritten

$$k_1[E][S] = (k_2 + k_3)[E:S]$$
$$k_1([E_o] - [E:S])[S] = (k_2 + k_3)[E:S]$$
$$[S][E_o] = \frac{(k_1 S + k_2 + k_3)[E:S]}{k_1}$$

letting
$$K_m = \frac{k_2 + k_3}{k_1}$$

then
$$[E:S]\frac{[S][E_o]}{K_m + [S]}.$$

With the k_4 process ignored in (H1), the specific rate constant for the decomposition of $E:S$ to enzyme and products is k_3. Defining velocity, V, as the rate of product formation for the enzymatic process, we obtain

$$V = k_3[E:S]$$

then
$$V = \frac{k_3[S][E_o]}{K_m = [S]}$$

in which K_m is the substrate concentration at which the reaction develops half its maximum velocity. The plot of reaction velocity V against substrate concentration $[S]$ is known as the Briggs-Haldane form and is shown in Fig. H3. At lower substrate concentrations the introduction of competitive inhibitor is seen to reduce the reaction velocity by effectively reducing enzyme concentration.

The preceding standard biochemical formulation of enzyme kinetics exhibits several limitations which limits its effectiveness in the study of enzyme kinetics. The quasi steady-state assumption presumes that $[E:S]$ reaches its steady value essentially instantaneously which precludes analytical treatment of the initial transients. Analysis is also virtually impossible for those combinations of rate constants which invalidate the assumption that $d/dt[E:S]$ is zero, and for the case when initial enzyme concentration is of the same order of magnitude as initial substrate concentration. The analog model presented here eliminates these limitations and further allows the study of competitively inhibited systems.

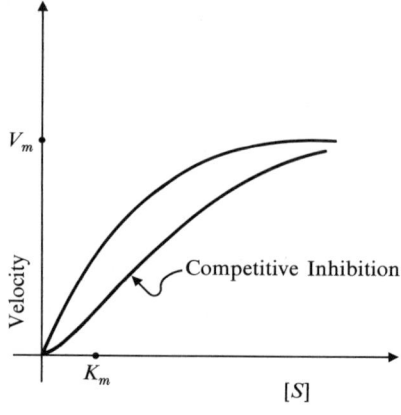

Fig. H3

Analog Model

Figure H4 shows the analog computer program which models Michaelis-Menten Enzyme Kinetics. The program is a direct representation of the six simultaneous first order differential equations developed above. Unity scaling is employed so that 1 machine unit corresponds to a normalized concentration value of 1.0. The time axis is scaled so that 1 second corresponds to the normalized rate of 1.00 (i.e., take k_1, k_2, or k_3 as the rate of change of normalized concentration $1/a$ sec. = 1.00. Then 1.00 sec. corresponds to a sec.). Switch 1 is included so that competitive inhibition may be separately investigated. If insufficient computer components are available the investigation of inhibition may be omitted without disturbing the remainder of the system. Problem variables are

P = Product Concentration
E = Enzyme Concentration
S = Substrate Concentration
$E:S$ = Enzyme-Substrate Complex Concentration
E_o = Initial Concentration Enzyme
S_o = Initial Substrate Concentration
V = Reaction velocity
$E:I$ = Enzyme-Inhibitor Complex Concentration.

Suggested Experimental Procedure

Prepare an analog computer patch panel to mechanize the program given in Fig. H4 and apply it to the computer. Set all attenuators as shown

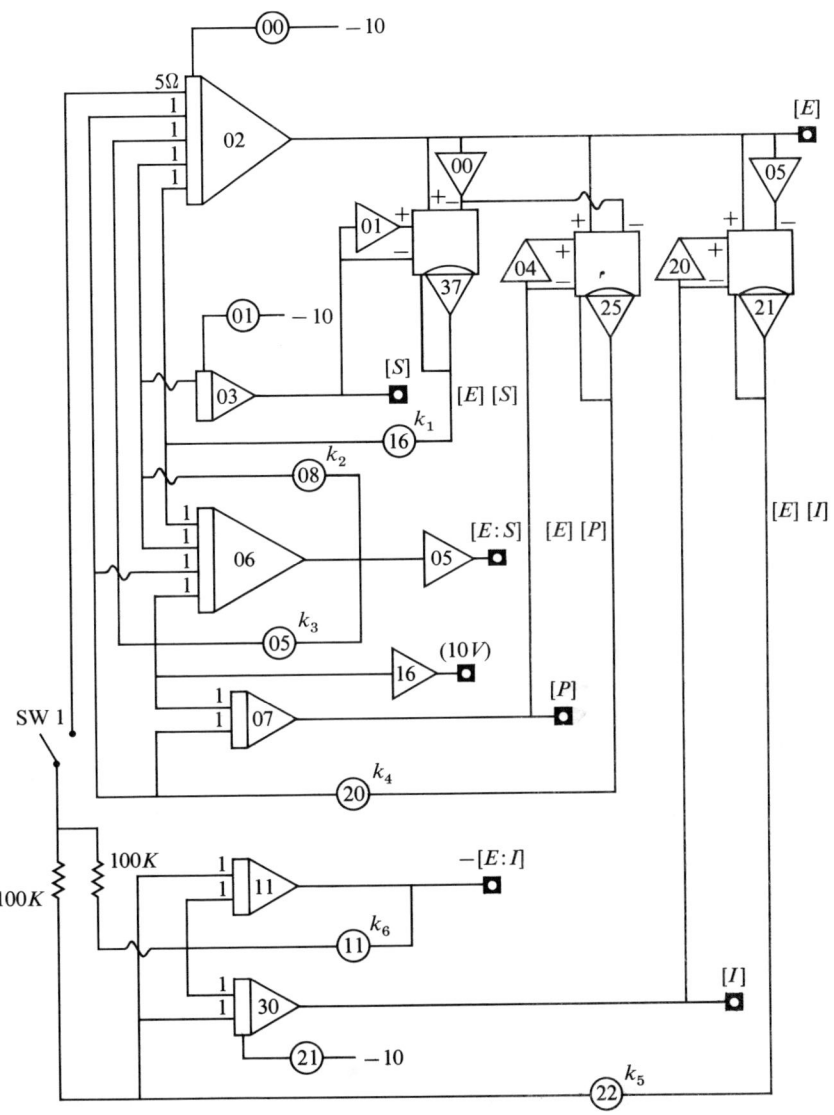

Fig. H4

in the potentiometer assignment sheet of Fig. H5 and perform a Static Check.

A. *Initial Concentration Ratios*

1. Set the rate constant attenuators as shown in Run 1 of the potentiometer assignment sheet.

2. With SW 1 open make three runs at E_o/S_o (Pot 00) ratios of 0.1, 0.3, 1.0. Record the time variation of [S], [P], [E:S], and [E].*

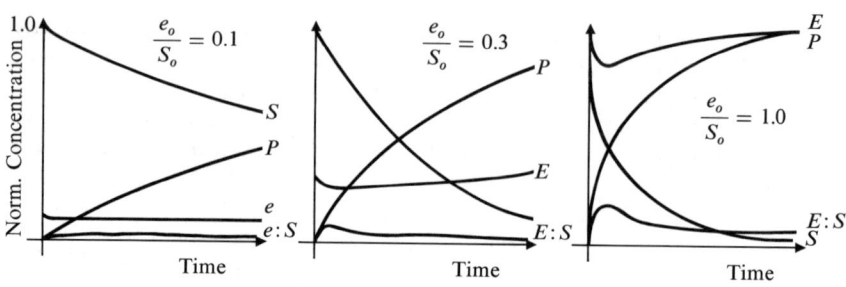

Fig. H5

3. With SW 1 closed and $E_o/S_o = 0.3$ make three runs at I_o/S_o (Pot 21) of 0.1, 0.3, 1.0. Record the concentration time variations as before.

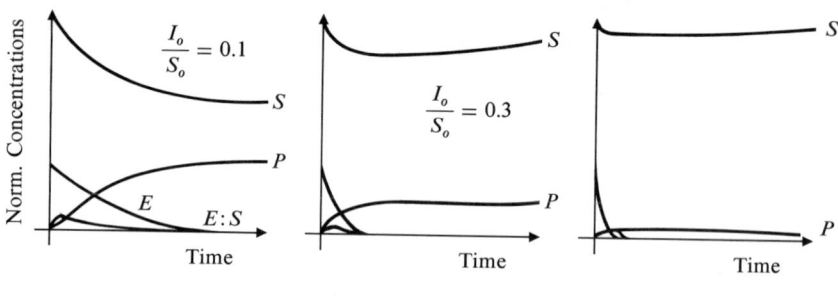

Fig. H6

B. *Briggs-Haldane Form*
1. Set attenuators as shown in Run 2 of the potentiometer assignment sheet.
2. With SW 1 open plot $10V$ vs. S (see Fig. H4) on the X-Y plotter, at three k_1 settings (Pot 16) of 0.2, 0.5, 1.0. Calculate k_m and V_{max} as shown in the mathematical formulation.
3. With SW 1 closed plot $10V$ vs. [S] as in B.2.

---

*These experiments are best performed in Repetitive Operation, the behaviors being viewed on an oscilloscope. If a computer with this mode is not available, then normal operation can be used, observations being made on an X-Y plotter or strip-chart recorder.

APPENDIX H. DYNAMIC BEHAVIOR OF ENZYME SYSTEMS

C. Reaction Rate Constants

1. With SW 1 open and all rate constants set as shown in Run 3 of the potentiometer assignment sheet make three runs at different values of k_1.
2. Set k_1 to original value in Run 3 and repeat for other rate constants. This procedure may be repeated until full familiarization has been achieved with the influence of each rate constant.
3. Repeat steps 1 and 2 under the influence of competitive inhibition. Notice that rate constants k_5 and k_6 have now been introduced to the system.

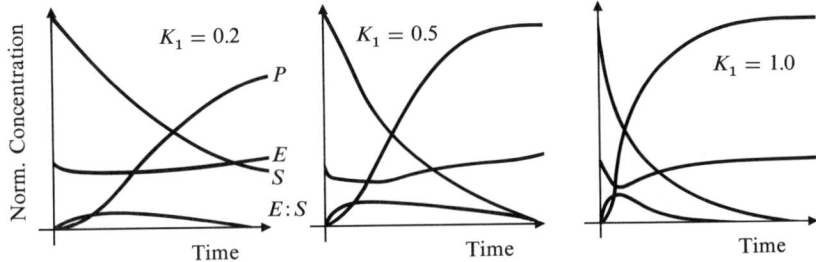

Fig. H7

POTENTIOMETER ASSIGNMENT SHEET

Date _____   Problem Dynamic Behavior of Enzyme Systems

Pot. No.	Parameter Description	Setting Static Check	Standard Setting	Setting Run No. 1	Setting Run No. 2	Setting Run No. 3	Notes
00	$[E_o]$	0.3		0.1		0.3	
01	$[S_o]$	1.0	1.0	1.0		1.0	
02							
03							
04							
05	k_3	0.1		1.0		1.0	
06							
07							
08	k_2	0.2		0.05		0.05	
09							
10	time base	0.02	0.02				
11	k_6	0.1		0.05		0.05	
12							
13							
14							
15							
16	k_1	1.0		0.2		0.2	
20	k_4	0.2		0.05		0.05	
21	$[I_o]$	0.1		0.1		0.1	
22	k_5	1.0		0.2		0.2	
23							
24							

AMPLIFIER ASSIGNMENT SHEET

Date _____   Problem __Dynamic Behavior of Enzyme Systems__

Amp. No.	FB	Output Variable	Static Check				Notes
			Calculated		Measured		
			Check Pt.	Output	Check Pt.	Output	
00							
01							
02		$[E]$					
03		$[S]$					
04							
05							
06							
07		$[P]$					
08							
09							
10							
11		$-[E:I]$					
12							
13							
14							
15							
16							
17							
18							
19							
21		$[E][I]$					
25		$[E][P]$					
30		$[I]$					
37		$[E][S]$					

APPENDIX I

R-L-C Transducer

TRANSDUCERS AND FILTERS HAVE an important role to play in communications and control systems work.

The filter indicated in Fig. I1 was used in a control system network to obtain a compensating effect. It is the purpose of this work to include an analysis of this network and a program which will enable one to simulate this filter on the analog computer.

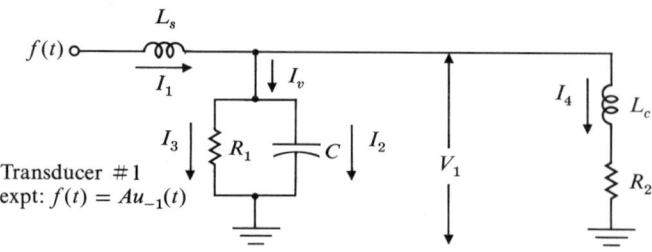

Fig. I1

The system equations are

$$V = L \frac{dI}{dt}$$

$$\frac{dI_1}{dt} = \frac{V_{L_s}}{L_s} = \frac{f(t) - V_1}{L_s} \tag{I1}$$

$$V = L \frac{dI}{dt}$$

$$\frac{dI_4}{dt} = \frac{V_{L_c}}{L_c} = \frac{V_1 - V_{R_2}}{L_c} = \frac{V_1 - I_4 R_2}{L_c} \tag{I2}$$

APPENDIX I. R-L-C TRANSDUCER

$$I_1 = I_v + I_4$$
$$I_v = I_1 - I_4 = I_3 + I_2$$
$$I_1 - I_4 = \frac{V_1}{R_1} + C\frac{dV_1}{dt}$$
$$\frac{dV_1}{dt} = \frac{I_1}{C} - \frac{I_4}{C} - \frac{V_1}{CR_1}. \tag{I3}$$

The parameters of the system are
$$L_s = 8\eta h = 8 \times 10^{-9} h$$
$$C = 13.7 pf = 13.7 \times 10^{-12} f$$
$$L_c = 2\eta h = 2 \times 10^{-9} h$$
$$R_2 = 25\Omega$$
$$R_1 = 3\Omega$$
$$f(t) = Au_{-1}(t) = 0.1\mu_{-1}(t) \quad \text{(step function)}$$

V_1 max will be estimated at twice the steady state value [A], therefore

$$V_{1\,\text{max}} = 0.2$$
$$I_{3\,\text{max}} = 0.2/R_1 = 0.2/3 \cong 0.1$$
$$I_{4\,\text{max}} = 0.2/R_2 = 0.2/2.5 \cong 0.01$$

therefore,

$$I_{1\,\text{max}} \cong I_{3\,\text{max}} + I_{4\,\text{max}}$$
$$= 0.1 + 0.01 \cong 0.1$$
$$\dot{I}_{1\,\text{max}} \cong \frac{|f(t)| + |V_1|}{L_s} = \frac{0.2}{8 \times 10^{-9}} = 0.025 \times 10^9$$
$$\dot{I}_{1\,\text{max}} \cong 2.5 \times 10^7 \cong 5 \times 10^7 \quad \text{(conservative)}$$
$$\dot{I}_{4\,\text{max}} \cong \frac{|V_{1\,\text{max}}|}{L_c} + \frac{|I_{4\,\text{max}}R_2|}{L_c}$$
$$\cong \frac{0.2}{2 \times 10^{-9}} + \frac{(0.01)(.25)}{2 \times 10^{-9}}$$
$$\cong 0.1 \times 10^9 + 0.125 \times 10^9$$
$$\cong 0.25 \times 10^9 = 25 \times 10^7$$
$$\dot{V}_{1\,\text{max}} \cong \frac{|I_{1\,\text{max}}|}{C} + \frac{|I_{4\,\text{max}}|}{C} + \frac{|V_{1\,\text{max}}|}{CR_1}$$
$$\cong \frac{0.1}{10^{-13}} + \frac{0.01}{10^{-13}} + \frac{0.2}{3 \times 10^{-3}}$$
$$\cong 0.01 \times 10^{12} + 0.001 \times 10^{12} + 0.01 \times 10^{12}$$
$$\cong 0.011 \times 10^{12} \cong 10^{10}.$$

So far, the given equations have been verified, the circuit was correct, the estimated maximum values have been calculated based on the step function $f(t) = Au_{-1}(t)$, where $A = 0.1$. The following table results:

Variable	Estimated Max	Scaled Variable
V_1	0.2	$[5V_1]$
$\dot{V}_1$	10^{10}	$[\dot{V}_1/10^{10}]$
I_1	0.1	$[10I_1]$
$\dot{I}_1$	5×10^7	$[\dot{I}_1/5 \times 10^7]$
I_4	0.01	$[100I_4]$
$\dot{I}_4$	25×10^7	$[\dot{I}_4/25 \times 10^7]$

$$\dot{I}_1 = \frac{f(t)}{L_s} - \frac{V_1}{L_s}$$

$$5 \times 10^7 \left[\frac{\dot{I}_1}{5 \times 10^7}\right] = \frac{[10f(t)]}{10L_s} - \frac{1}{5L_s}[5V_1]$$

$$\left[\frac{\dot{I}_1}{5 \times 10^7}\right] = \frac{[10f(t)]}{5 \times 10^8 L_s} - \frac{1}{25 \times 10^7 L_s}[5V_1] \qquad (14)$$

$$\dot{I}_4 = \frac{V_1}{L_c} - \frac{I_4 R_2}{L_c}$$

$$\left[\frac{\dot{I}_4}{25 \times 10^7}\right] = \frac{[5V_1]}{125 \times 10^7 L_c} - \frac{R_2}{25 \times 10^9 L_c}[100I_4] \qquad (15)$$

$$\dot{V}_1 = \frac{I_1}{C} - \frac{I_4}{C} - \frac{V_1}{CR_1}$$

$$\left[\frac{\dot{V}_1}{10^{10}}\right] = \frac{[10I_1]}{10^{11}C} - \frac{[100I_4]}{10^{12}C} - \frac{[5V_1]}{5 \times 10^{10}CR_1} \qquad (16)$$

The scaled (normalized) Eqs. are (I4), (I5), and (I6).
The suggested diagram follows:

$$\boxed{\beta = 10^{10}}$$

The analysis considered on the previous pages is correct for all of the assumptions made. However, the circuit did not work when it was put on the computer. Rescale the equation for:

$$|\dot{V}_{\max}| = 10^8$$

and

$$|\dot{I}_4| = 25 \times 10^5$$

After a complete rescaling has been done, put this problem on the computer (TR-20) and verify that it works properly.

It is expected that the student will apply various input functions to the simulated *filter* to determine its behavior.

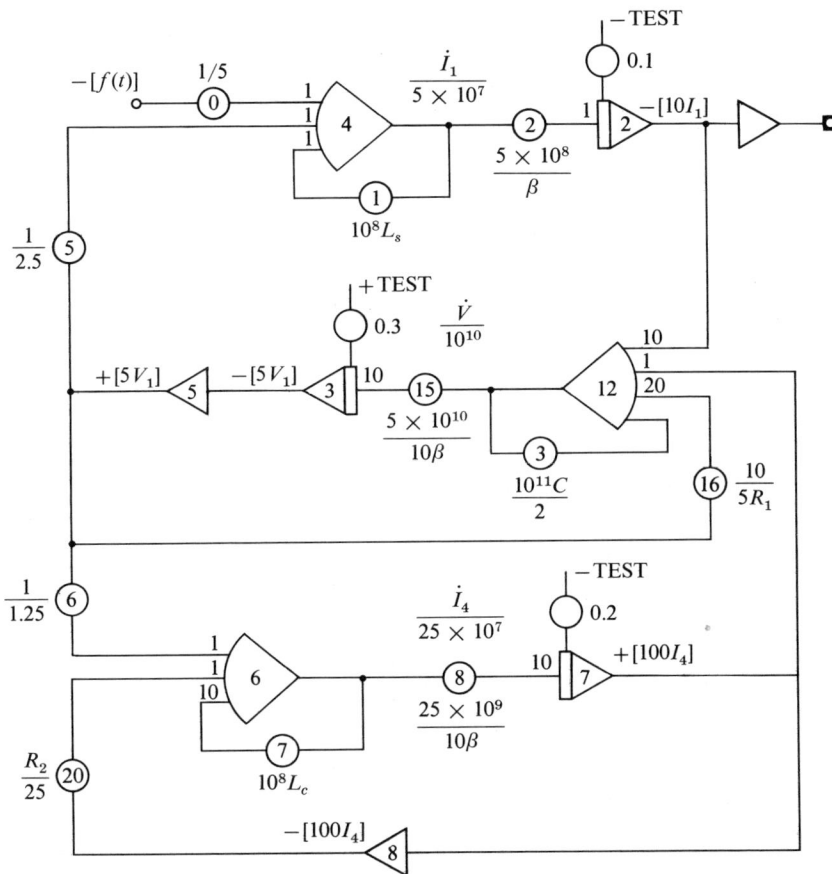

Fig. I2. Scaled Diagram

POTENTIOMETER ASSIGNMENT SHEET

Date _____   Problem ___R-L-C Transducer___

Pot. No.	Parameter Description	Setting Static Check	Static Check Output Voltage	Setting Run Number 1	Notes	Pot. No.
00	$1/5$	0.2		0.2		
01	$10^8 L_s$	0.8		0.8		
02	$5 \times 10^8/\beta$	0.05		0.05		
03	$10^{11} C/2$	0.685		0.685		
04						
05	$1/2.5$	0.400		0.400		
06	$1/1.25$	0.800		0.800		
07	$10^8 L_c$	0.200		0.200		
08	$25 \times 10^9/10\beta$	0.2500		0.2500		
09						
10						
11						
12						
13						
14						
15	$5 \times 10^{10}/10\beta$	0.5		0.5		
16	$10/5R_1$	0.667		0.667		
17						
18						
19						
20	$R_2/25$	1.000		1.000		
21						
22						
23						
24						

APPENDIX I. R-L-C TRANSDUCER 225

AMPLIFIER ASSIGNMENT SHEET

Date _____   Problem __R-L-C Transducer__

Amp. No.	FB	Output Variable	Static Check				Notes
			Calculated		Measured		
			Check Pt.	Output	Check Pt.	Output	
00							
01							
02	$\int$	$-[10I_1]$	$+0.0050$	$+0.1000$			
03	$\int$	$-[5V_1]$	$+0.840$	-0.300			
04	H.G.	$[\dot{I}_1/5 \times 10^7]$		$+0.1000$			
05	I	$+[5V_1]$		$+0.3000$			
06	H.G.	$-[\dot{I}_4/25 \times 10^7]$		-0.0200			
07	$\int$	$+[100I_4]$	-0.2875	$+0.200$			
08	I	$-[100I_4]$		-0.200			
09							
10							
11							
12	H.G.	$[\dot{V}_1/10^{10}]$		-0.1680			
13							
14							
15							
16							
17							
18							
19							
20							

APPENDIX J

Estimation of Maximum Values from Two Simultaneous Differential Equations

Introduction

One of the problems which an analog computer programmer encounters is that of estimating the maximum values of the outputs of amplifiers. At times this is an easy problem while at other times it is more difficult. This summary is intended to present a technique for estimating the maximum value of a variable and its derivative given the maximum value of a second variable in a differential equation.

Estimation of the Maximum Value

If one encounters the following two differential equations (which are the equations of motion of a libration point between the earth and moon) or equations somewhat like these, and has the problem of estimating the maximum value of $\dot{y}$, $\dot{x}$ and x given the maximum value of $y(y_{max})$ the following procedure will prove helpful:

Given:

$$\ddot{x} = +2\dot{y} + 3/4\,x + 3/4\sqrt{3}(1 - m^*)y \tag{J1}$$
$$\ddot{y} = -2\dot{x} + 9/4\,y + 3/4\sqrt{3}(1 - m^*)x \tag{J2}$$

where y_{max} is known, calculate x_{max}, $\dot{x}_{max}$, and $\dot{y}_{max}$.

$$m^* = \frac{m\ \text{moon}}{m\ \text{earth}} = 0.0121$$

$$m = \text{mass}$$

Solution

The solution to this problem becomes clear when one recalls that at the maximum value of a function the first derivative vanishes and the second derivative is less than zero, that is,

at
$$y = y_{max}$$
$$\dot{y} = 0$$
and
$$\ddot{y} < 0$$

also,

at
$$x = x_{max}$$
$$\dot{x} = 0$$
and
$$\ddot{x} < 0.$$

Using these results, Eqs. (J1) and (J2) become:

$$3/4 x_{max} + 3/4\sqrt{3}(1 - m^*)y_{max} < 0 \qquad (J3)$$
$$9/4 y_{max} + 3/4\sqrt{3}(1 - m^*)x_{max} < 0. \qquad (J4)$$

These equations can also be written in the following way:

$$x_{max} < -\sqrt{3}(1 - m^*)y_{max} \qquad (J5)$$
and
$$3 y_{max} < -\sqrt{3}(1 - m^*)x_{max} \qquad (J6)$$
or
$$\frac{-3 y_{max}}{\sqrt{3}(1 - m^*)} < x_{max} < -\sqrt{3}(1 - m^*)y_{max}. \qquad (J7)$$

This last equation puts a very definite limit on x_{max}. That is, for some given y_{max} one could now estimate x_{max}.

As an example consider $y_{max} = 100$. Then,

$$\frac{-3(100)}{\sqrt{3}(1 - 0.0121)} < x_{max} < -\sqrt{3}(1 - 0.0121)(100)$$
$$-177.0 < x_{max} < -171.0.$$

Therefore a maximum value for x_{max} could be $|x_{max}| = 175$.

Now that y_{max} and x_{max} have been determined, one could estimate $\dot{y}_{max}$ and $\dot{x}_{max}$ from knowing the fact that

when
$$\ddot{y} \text{ and } \ddot{x} = 0$$
$$\dot{y} \text{ and } \dot{x} \text{ are maximum.}$$

Under these circumstances Eqs. (J1) and (J2) become

$$0 = +2\dot{y}_{max} + 3/4 x_{max} + 3/4\sqrt{3}(1 - m^*)y_{max} \qquad (J8)$$
$$0 = -2\dot{x}_{max} + 9/4 y_{max} + 3/4\sqrt{3}(1 - m^*)x_{max}. \qquad (J9)$$

Therefore

$$\dot{y}_{max} = -3/8 x_{max} + 3/8\sqrt{3}(1 - m^*)y_{max} \qquad (J10)$$

and
$$\dot{x}_{\max} = +9/8 y_{\max} + 3/8\sqrt{3}(1 - m^*)x_{\max}. \quad (J11)$$

From the values of $y_{\max}$ and $x_{\max}$ obtained previously:

$$|\dot{y}_{\max}| = +3/8(175) + 3/8\sqrt{3}(1 - 0.0121)(100)$$
$$|\dot{y}_{\max}| = 129.6$$
$$|\dot{x}_{\max}| = 9/8(100) + 3/8\sqrt{3}(1 - 0.0121)(175)$$
$$|\dot{x}_{\max}| = 226.15.$$

The table for the maximum values of the variables is indicated below. Note that the estimated maximum values have been changed to the rounded up maximum values (R.U.M.). That is, the numbers should be of the form

$$1, 2, 4, 5 \times 10^{\pm n}$$

where $n = 0, 1, 2, 3 \ldots$

(All maximums except the one for x should be rounded up.)

Problem Variable	Rounded-up Maximum	Computer Variable
y	100	$[y/100]$
x	175	$[x/175]$
$\dot{y}$	200	$[\dot{y}/200]$
$\dot{x}$	400	$[\dot{x}/400]$

It is hoped that this summary will be an aid to programmers wishing to perform scaling of variables involving simultaneous differential equations.

INDEX

Accumulator, 122
Alveolar partial pressure, 188
Amplifier control, 100–101
Amplitude scaling, 31–37
  rounded up maximum, 36
Analog computers, 1, 5, 51
Analog/Logic, hybrid, 2, 51
AND gate, 106
Automobile suspension system, 27–29, 131
Briggs-Haldane, 216
Cardiac output, 188
Check amplifier, 21
Chemical systems, 210
CO_2 rebreathing system, 187–197
Comparators, 91, 101
Competitive inhibition, 212
Counters, 116–120
D/A relay, 102, 103
D/A switch, 102
Differentiator, 111–113
Diode switches, 79
Electric fields, 160–162
Electron ballistics, 160–172
Enzyme systems, 210–219
Estimation, 226
Filters, 220
Flip-flop (bistable multivibrator), 109, 110
Function generators, 68–79
Hybrid program (bouncing ball), 112, 113
Inhibit, 108
Initial conditions, 47, 99
Integrator, 12, 13
Interfacing components, 97–99
Inverter (logic), 105
Kinetics, 198–209
Libration point, 226

Limiters, 80, 81
  feedback, 87
Linear systems, 51–54
Logarithmic generators, 78, 79
Logic circuits, 107, 108
Logic signals, 105
Mathieu's equation, 120, 121
Maximum values, 56–63, 226
Mode control, 17–20
Multivibrator (flip-flop), 107, 108
Non-linear programming, 65–95
  multiplication and division, 65–68
  function generation, 68–73
Operational amplifier, 9–11
OR gate, 106, 107
Parallel logic, 104–123
Physical systems, 1, 2
Potentiometer, 6–8, 79–85
Programming analog computers, 22–27
Programming symbols, 14–16
Protein chains, 210
Reaction rate constants, 217
Register circuits, 119
Registers, 114–116
Resistive networks, 70–73
R-L-C transducer, 220–225
Shift registers, 114–116
Sine generators, 74–79
Spring mass damper, 129–131
Static check, 43–47, 143
  Program check, 46, 47, 142, 143
  circuit, 43–44, 143
Tapered nozzle, 147, 159
Time scaling, 37–40
Timer, 122
Track/store, 99, 100
Tubular chemical reactor, 135–146
Venous partial pressure, 188